International
REVIEW OF
Neurobiology
Volume 76

International

REVIEW OF

Neurobiology

Volume 76

Temporal Lobe Epilepsy and the Mind–Brain Relationship: A New Perspective

EDITED BY

SHIRLEY M. FERGUSON

Department of Psychiatry, University of Toledo College of Medicine
(formerly Medical College of Ohio) Toledo, Ohio, USA

MARK RAYPORT

Department of Neurological Surgery and Radiology
Medical College of Ohio, Toledo, Ohio, USA

CAROLYN A. SCHELL

Comprehensive Epilepsy Program
Medical College of Ohio, Toledo, Ohio, USA

The Medical College of Ohio was founded in 1964. In 2005, the name of the institution was changed to Medical University of Ohio. On July 1, 2006, the Medical University was merged with the University of Toledo. The current name is University of Toledo College of Medicine.

ELSEVIER

AMSTERDAM • BOSTON • HEIDELBERG • LONDON
NEW YORK • OXFORD • PARIS • SAN DIEGO
SAN FRANCISCO • SINGAPORE • SYDNEY • TOKYO

Academic Press is an imprint of Elsevier

Academic Press is an imprint of Elsevier
525 B Street, Suite 1900, San Diego, California 92101-4495, USA
84 Theobald's Road, London WC1X 8RR, UK

This book is printed on acid-free paper. ♾

For information on all Academic Press publications
visit our Web site at www.books.elsevier.com

ISBN-13: 978-0-12-373667-3
ISBN-10: 0-12-373667-6

Printed and bound in the United Kingdom

Transferred to Digital Printing, 2010

DEDICATION

This book, "Temporal Lobe Epilepsy and the Mind–Brain Relationship: A New Perspective," is dedicated to the memory of Mark Rayport M.D. C.M., Ph.D., F.A.C.S., who died on March 4, 2003. The chapters of this book comprise some of the unpublished manuscripts written with his wife and collaborator, neuropsychiatrist, Shirley M. Ferguson, M.D., during his lifetime, and some written after Dr. Rayport's death.

The findings reflect Dr. Rayport's compelling dedication to and deep respect for the brain; his informed, skilled neurosurgery; his determination to achieve complete cessation of the seizures, and to gain understanding of and make a change in the lives of his patients. In essence, the book represents Dr. Rayport's interdisciplinary thinking, which not only aimed at accurate diagnosis of the epileptic focus, but at understanding the interplay between brain function and behavior.

CONTENTS

SECTION II
STEREOTAXIC STUDIES

Olfactory Gustatory Responses Evoked by Electrical Stimulation of Amygdalar Region in Man Are Qualitatively Modifiable by Interview Content: Case Report and Review

Mark Rayport, Sepehr Sani, and Shirley M. Ferguson

SECTION III
CONTROVERSY IN DEFINITION OF BEHAVIORAL DISTURBANCE

Pathogenesis of Psychosis in Epilepsy. The "Seesaw" Theory: Myth or Reality?

Shirley M. Ferguson and Mark Rayport

SECTION IV
OUTCOME OF TEMPORAL LOBECTOMY

Memory Function After Temporal Lobectomy for Seizure Control: A Comparative Neuropsychiatric and Neuropsychological Study

Shirley M. Ferguson, A. John McSweeny, and Mark Rayport

Life After Surgery for Temporolimbic Seizures

SHIRLEY M. FERGUSON, MARK RAYPORT, AND CAROLYN A. SCHELL

Appendix I

In Memoriam: Mark Rayport, M.D. C.M., Ph.D., F.A.C.S

Appendix II: Conceptual Foundations of Studies of Patients Undergoing Temporal Lobe Surgery for Seizure Control

MARK RAYPORT

CONTRIBUTORS

Numbers in parentheses indicate the pages on which the authors' contributions begin.

Shirley M. Ferguson, M.D. (3, 21, 35, 45, 65, 87), Professor Emerita, Department of Psychiatry. Neuropsychiatrist, Comprehensive Epilepsy Program. University of Toledo College of Medicine (formerly Medical College of Ohio), Toledo, Ohio 43614, USA

A. John McSweeny, Ph.D. (65), Professor, Department of Psychiatry and Department of Neurology. Neuropsychologist, Comprehensive Epilepsy Program. University of Toledo College of Medicine (formerly Medical College of Ohio), Toledo, Ohio 43614, USA

***Mark Rayport**, M.D. C.M., Ph.D., F.A.C.S (3, 21, 35, 45, 65, 87, 117, 123), Professor Emeritus, Department of Neurological Surgery and Department of Radiology. Founding Chairman of the Department of Neurological Surgery. Director and Neurosurgeon, Comprehensive Epilepsy Program. Medical College of Ohio, Toledo, Ohio 43614, USA

Sepehr Sani, M.D. (35), Resident, Department of Neurological Surgery, Rush Presbyterian-Saint. Luke's Medical Center, Chicago, Illinois 60612, USA, Graduate of the Medical College of Ohio

†**Carolyn A. Schell**, B.S.N., R.N. (87), Nurse Coordinator, Comprehensive Epilepsy Program (1980–2003), Medical College of Ohio, Toledo, Ohio 43614, USA

*Deceased March 4, 2003.

†Retired October 1, 2003.

ACKNOWLEDGMENTS

The inspiration for this volume was Dr. Rayport's, and although he could not participate in the final editing, he has been with us as a guiding spirit throughout. This volume is dedicated to his memory.

The contribution of Carolyn A. Schell, R.N., B.S.N., to this volume is greatly appreciated for her outstanding insights into the understanding and care of the epilepsy surgery patient, her remarkable memory for data about the patients, and her impressive skills in helping to prepare this manuscript.

The generous and most informed assistance of Andres M. Kanner, M.D., in assessing and preparing the material for publication was invaluable.

For their encouragement and their keen and insightful comments and suggestions along the way, I wish to thank our children: Stephen Rayport, M.D., Ph.D.; Jeffrey Rayport, Ph.D.; and Jennifer Rayport Rabodzeenko, Ph.D.

I am most appreciative of and touched by the front cover artwork created by our son-in-law, Andrei Rabodzeenko, who was inspired to make this drawing after a conversation with Dr. Rayport about brain function.

Shirley M. Ferguson

INTRODUCTION

This book, devoted to the surgical treatment of epilepsy and its consequences, provides an extraordinary perspective on the fascinating question of the relationship between brain and mind. Contrary to the current emphasis on statistics and objectivity, this book is dedicated to understanding the whole person, the life and experiences of the individual. It reports on cutting-edge technical skills provided with a human touch. This book grew out of collaboration between (the late) Mark Rayport, M.D. C.M., Ph.D., Professor Emeritus of Neurological Surgery and Radiology at the Medical College of Ohio (now University of Toledo College of Medicine), who died in March 2003, and myself, a neuropsychiatrist before the formal establishment of the subspecialty. I collaborated with Dr. Rayport following our training at McGill University. Dr. Rayport completed his residency with Wilder Penfield, M.D., at the Montreal Neurological Institute and I completed my residency in psychiatry with D. Ewen Cameron, M.D., at the Allan Memorial Institute of Psychiatry in the 1950s. We collaborated from the 1960s, first at the Albert Einstein College of Medicine and, from the 1970s, at the Medical College of Ohio, until our retirement in 1993.

Over the course of this collaboration of almost 50 years, we followed more than 90 temporolimbic epilepsy patients prospectively in an interdisciplinary manner from the time of the initial consultation, through trial of medication, moving then to incisive preoperative diagnostic studies (including implantation of depth electrodes and brain stimulation, if indicated), surgery under local anesthesia (with brain stimulation), to postoperative course, in some cases (at the Medical College of Ohio) for over 20 years. From this experience, we gleaned major insights into the interactions of brain and mind.

This book reflects the understanding that the temporolimbic seizure originates in tissue providing the substrate for the emotional life and memory. This provides a fascinating opportunity to open a magic window on thinking and behavior since the subjective experience can be communicated and the behavior can be observed. In the book chapters, we show that results of brain discharge, occurring spontaneously (during a seizure) or in response to electrical brain stimulation, may furnish unique responses which relate to the patient's life experiences. One can, thus, be present at the true brain-mind interface. Such data shed light on the physiological underpinnings of Freud's theories of ego structure. They also provide evidence that context can influence the quality of the brain response and that psychosis occurs in subtle conjunction with seizure discharge, not, as has been suggested, as an alternative in a "seesaw" relationship. In the cognitive area, memory function may improve after successful surgery (when the noxious abnormal electrical discharge is eliminated) despite dire predictions of the negative effects of brain ablation. Temporal lobectomy for intractable seizures can produce freedom from seizures and a most impressive, positive change in the patient's life. Such a result is attributable to the highly informed skill of the neurosurgeon and to the exceedingly productive effect of the collaboration between neurosurgeon and neuropsychiatrist throughout the patient's course.

Although we published widely in refereed journals and book chapters and made presentations at courses, and at national and international meetings, we never pulled together our experience in a single volume. Some of the work was not published and, was, indeed, in progress at the time of Dr. Rayport's death, and is now being completed.

I would call attention to the time frame of the chapters in this volume. The work represents our thinking, and its evolution, about important issues pertaining to mind-brain relationships acquired through observations made during the study and treatment of patients with intractable temporal lobe epilepsy dating from 1970 to the present.

Different aspects of some cases may be discussed in different chapters. Material about some patients appears more than once, as different aspects of their behavior are discussed. In the first chapter, patients are cited by case number. In subsequent chapters, patients have been assigned acronyms, not their initials.

Shirley M. Ferguson

SECTION I

PHYSIOLOGICAL CORRELATES OF FREUD'S THEORIES

THE ID, THE EGO, AND THE TEMPORAL LOBE

Shirley M. Ferguson* and Mark Rayport[†,‡,1]

*Department of Psychiatry, University of Toledo College of Medicine (formerly Medical College of Ohio), Toledo, Ohio 43614, USA

[†]Department of Neurological Surgery, Medical College of Ohio, Toledo, Ohio 43614, USA

[‡]Department of Radiology, Medical College of Ohio, Toledo, Ohio 43614, USA

During our first 10 years of interdisciplinary study of patients with intractable temporal lobe epilepsy at the Albert Einstein College of Medicine, we were struck again and again by the resurgence in the seizure of mental content reflecting unconscious material. The similarity with cases described by Freud made us posit that during the seizure, loss of ego control may lead to resurgence of unconscious material. Thus, the seizure appeared to provide the physiological substrate for Freud's theory of ego structure.

This paper was presented to a medical group in Toledo, Ohio, when we arrived in 1969 to join the faculty of the newly established Medical College of Ohio.

I. Introduction

It may be noted in the writings of some of the great figures in neurology–epileptology, a poetic, sometimes pastoral, quality, as well as great humility.

[1]Deceased.

INTERNATIONAL REVIEW OF NEUROBIOLOGY, VOL. 76
DOI: 10.1016/S0074-7742(06)76001-9

Among them, Hughlings Jackson, Wilder Penfield, and William Lennox express in their writings the great privilege of having the opportunity to witness and affect basic workings of the mind, to deal with what appears as the physical substrate of human behavior. In the nineteenth century, Jackson (1875) made classical observations of the phenomena of epilepsy. He suggested that, "Understanding of the dreamy states of temporal lobe epilepsy might be the key to the understanding of insanity." Freud (1953) stated that, "The interpretation of dreams is the royal road to a knowledge of the unconscious activities of the mind." Penfield (1938) spoke of epilepsy as an experiment of nature from which many things could be learned about human behavior. In 1948, Frederick Gibbs (1951) wittily described the sylvian fissure as marking the border between neurology and psychiatry. Ernst Rodin (personal communication), a neurologist working in this field, has spoken of patients with temporal lobe seizures as presenting the id upon a platter. Robert Doty (1967), a neurophysiologist, believes that the psychiatrist of the future might be a neurologist specializing in disorders of the limbic system.

The present observations and evidence come from neurological, neurophysiological, neurosurgical, and behavioral studies of patients with temporal lobe epilepsy. The thesis which we will attempt to develop is that in the temporal lobe region of the brain may well lie anatomical structures with the physiological mechanisms necessary for the framework of ego structure as formulated by Freud in behavioral terms. The initial task then will be to provide orientation to the question of cerebral localization, to describe the function of the area of the brain involved, and to analyze behavioral phenomena of temporal lobe seizures.

There are those who have argued that the brain is a homogeneous mass and that function is impaired only in relation to the quantity of brain tissue damaged. Others, the so-called localizers, hold quite opposite viewpoints. It is only recently, with improvement in and development of new techniques and instruments, that there is an increased chance to probe this question of localization more accurately.

Historically, in the Middle Ages and Renaissance, the main concern was to localize the seat of the soul. At the beginning of the nineteenth century, phrenology was introduced by Gall and Spurzheim (Freedman *et al.*, 1976). This marked the beginning of the controversy between partisans of localization and the antilocalizers. However, it was not until 1861, with the work of Broca (1861), that the first datum appeared showing a definite relationship between a lesion in a particular brain area and a change in function. Clinical phenomena were looked at more closely and by penetrating description, other disorders of higher functions were teased out and assigned a cerebral localization. In 1891, Freud (Moruzzi and Magoun, 1949) proposed the term agnosia or disorder of recognition. He also wrote a monograph on aphasia (Freud, 1891). Excess in terms of localization was followed by a reaction best exemplified by the Gestalt School which led to the notion of a general intellectual disability, defined for instance by

Goldstein (1995) as a loss of abstract ability. In 1929, Lashley (Heilman and Valenstein, 1993) suggested that cerebral organization depended on the participation in function of large masses of nervous tissue and its impairment as directly proportional to the mass of nervous tissue destroyed, irrespective of localization. At the very same time, such localizers as Kleist (Weisenburg and McBride, 1935) and Nielsen (Nielsen, 1936) were holding forth in attempting to assign a special function to each of the cytoarchitectonic areas in the cerebral cortex, which anatomists of the schools of Brodmann, Von Economo, and the Vogts had described.

In contrast to most of the great systems of the neocortex which are concerned with the external world, the limbic system is concerned with the self. It controls the exchange between the body and the external world and, through its command over the autonomic and endocrine systems, regulates the internal world. As a repository of such significant behavior as feeding, drinking, defense, aggression, sex, and maternal responses, associated stimuli are motivational. The limbic system assigns motivational value to particular stimuli. Animal ablation work provides evidence for such statements. For example, removing all the neocortex of a cat produces no personality change, whereas lesions placed in the ventral medial nucleus of the hypothalamus change the behavior of the animal from gentleness to ferociousness. Bilateral amygdalectomy causes savageness to disappear.

In the 1950s, a widely held compromise position was popularized in neuropsychological circles by the Russian neuropsychologist, Luria (1966), a pupil of Vygotsky. He describes the higher functions as complex, and with wide representation, but with localizable contributions—thus the variations and the type of impairment of a function depend on the position in the circuit where the lesion occurs. It is this kind of localization, not of whole functions, but of contributions to function which is the most *au courant* notion and modification of the theory of cerebral localization.

There are certain types of cerebral localization of function which are now accepted without question such as the motor projection and sensory receptor zones; the pre- and postcentral areas; the seat of consciousness in the subcortical formation which resulted from the work of Magoun (Moruzzi and Magoun, 1949) on the ascending reticular formation; and Penfield's theory (Penfield, 1938) of the centrencephalic system. Localization of the emotional life is illustrated by the work of Klüver and Bucy (1937, 1939) who did removals of major portions of the temporal lobes and rhinencephalon, including the amygdala, uncus, hippocampus, and hippocampal gyrus and have shown striking alterations in behavior. These operations were done on mature rhesus monkeys. Lesions produced visual agnosia, loss of ability to detect the meaning of animate or inanimate objects on the basis of visual criteria, and a tendency to examine all objects orally. There was a strong tendency to take notice of every visual stimulus.

There were changes in emotional behavior with absence or decrease in emotional responses, which might reflect an unmodulated response showing impaired judgment. The id would thus appear?

An essential ingredient for intact ego function is the retention of ongoing experience. As far back as 1898, Von Bechterew (Zola-Morgan *et al.*, 1986) described bilateral lesions of the hippocampus in a patient with a severe amnesia of registration. The pertinent work in this area was done by Penfield and Milner (1958) and Scoville and Milner (1957) who described characteristic, incapacitating retention deficits in patients with hippocampal ablations. The lateral temporal cortex appeared to be essential for proper ego function. Penfield (Penfield and Jasper, 1954) designated the lateral temporal cortex as "the brain tissue of comparison." His stimulating electrode during epilepsy surgery might reproduce seizure phenomena in which perceptions, whether simple or complex, were altered: The environment would become more or less familiar, or past experience would be relived.

II. Phenomenology of Seizures

A clear understanding of the nature and variations of the seizure elements is essential for hypothesizing about behavioral components.

Seizures may be classified as generalized or focal in origin. It is necessary at the outset to caution against too ready classification. Because a patient reports generalized convulsions, one must not ignore the possibility that focal seizures may generalize. In fact, 50% of focal seizures do just that. Among focal seizures, the most common site is the temporal lobe.

The temporal lobe focal seizure is unique in form and in content. While the grand mal attack is a circumscribed event starting at an identifiable moment, associated with complete absence of consciousness, and having definite duration, the temporal lobe seizure may be heralded by (1) a prodromal period lasting for hours or days; (2) an aura with clear or subtle changes in consciousness and reality contact; (3) an ictus characterized by loss of contact which may be associated with simple or complex automatic behavior; and (4) a postictal period of varying lengths, possibly with confusion, thinking difficulty, mood change, and/or amnesia. More primitive manifestations, such as an epigastric sensation and automatic movements like those of mastication, may be identical from patient to patient. Often, however, the seizure experience is highly personal—for the very reason that we are dealing with tissues which belong to the limbic system and are those involved with the organization of emotion and behavior, with time sense and sequencing, memory, speech and are, thus, important tissues of the psychic life. As Hughlings Jackson (1875) said, "What form a man's mental automatism will take depends, I think, very much on what his natural disposition is." In other words, the content of the seizure will be woven out of the threads of the patient's life—past and present.

The psychical phenomena occurring during temporal lobe seizures will depend on the locus of the epileptic focus. There may be perceptual changes: In the visual sphere, objects may look larger or smaller; there may be distortions: The floor and walls may appear wavy. In the auditory sphere, sounds may appear to be more distant or louder or confused. Words of others may become unintelligible, as if they were speaking in a foreign tongue. There may be changes in self-awareness with depersonalization, derealization, *déjà vu,* or *jamais vu* sensations. The patient may experience forced thought of a word, a sentence, obscene words, nonsense words. One of our patients came repeatedly with a piece of paper on which there was written what he called "gibberish." After a seizure, he would immediately record what he had heard, because it was always the same, and it seemed to have meaning, but he could never capture it. The patient may have an "experiential seizure," described by Penfield (Penfield and Jasper, 1954) as an exact reproduction of vivid past experiences which would come to the patient in full emotional context, as if a needle were put on the groove of a record. At the same time, the patient was aware of his present location. Later, Penfield (Penfield and Perot, 1963) modified this concept by recognizing that the experiential response could be related also to ongoing events in the patient's life. The patient may experience changes in mood with fear, depression, elation, and religiosity. During complex hallucinatory experiences, the patient may hear his name, music, and so on. He may have an olfactory experience which usually is unpleasant.

This seems like very dramatic symptomatology. However, ictal events arising from the temporal lobe may be subtle, sometimes not observed at all even by those close to the patient, or misinterpreted as bad behavior. The seizure may vary greatly depending on which portion or portions of the temporal area are involved. It may vary in complexity and severity and may be largely subjective and fleeting. Rather than a definite, clear-cut seizure, there may be a change in perception of the setting, of the patient's physical surroundings, in what he sees or hears, or there may be a sudden shift in the level of total functioning with abrupt derailment of the associative thread. There may be a breakdown in integrated behavior with suppressed or repressed feelings emerging in motor and/or verbal behavior, as if a short circuiting has occurred—the censor mechanism and reality-oriented functions in the ego structure having failed. Thus, the id may dominate.

It may be appropriate here to call attention again to the popular thesis that individuals with epilepsy are normal except for the susceptibility to occasional bursts of electrical discharge from their brains. Historical figures are held up as examples of excellent functioning despite the occurrence of seizures. One may as well point to such equally eminent people as Dostoevski and Van Gogh for evidence of the variation in experiencing and behavioral reaction, sometimes positive and welcomed, and, sometimes, extremely negative, feared, and destructive in those whose epileptic discharges involve the temporal lobe region. One may recall the claim in the literature that epileptics tend to religiosity. Dostoevski (Rice, 1985), for instance, wrote, "The air was filled with a big noise and I thought that it had engulfed me. I have really touched God. He came into myself. 'Yes, God exists,' I cried and I don't remember anything else. You all healthy people can't imagine the happiness which we epileptics feel the second before our fit. Mohammed said that he had seen paradise and had gone into it. He did not lie. He had been in paradise during an attack of epilepsy. He was a victim of this disease like I was. For all the joys that life may bring, I would not exchange this one." As evidenced in his paintings, Van Gogh probably experienced great exuberance, but, also, had episodes of violence such as required his institutionalization.

Among patients with seizures in the temporal lobe, studies have repeatedly indicated a high incidence of psychiatric disorder. Ostow (1954) said that one might encounter "the whole gamut of psychiatric disorder in these patients." How do we account for this? One approach to this problem is longitudinal investigation of each patient in terms of his psychodynamics, his seizure experiences and their relation to his ongoing life situation, and the state of his higher cerebral functions concurrent with neurosurgical studies and procedures. The data presented come from study of the ictal state, the so-called interictal period and psychotic episodes. Various compartments of the patient's experience may be separated for consideration.

III. Clinical Findings in Patients Treated at the Albert Einstein College of Medicine, Bronx, New York, During the 1960s

Evidence for our hypotheses has been taken from findings in 18 patients (in our series of 45 patients) who sought surgical treatment for relief of intractable temporal lobe seizures. Their experiences and responses may be likened to a key or a mirror reflecting brain organization and brain–behavior relationships.

A. Positive Experiences

Case 1, a policeman, described greatly enhanced visual and auditory experiences during his aura: His surroundings and objects became extremely beautiful; music was lovelier than ever before. At other times, he would become acutely paranoid and homicidal.

Case 2, a scientist would always have the feeling in his seizure that he was on the verge of a great discovery, that his thesis was about to jell, and he would regret that the seizure did not last quite long enough.

Case 3, a young doctor, also an enthusiastic tennis player, had an illusionary aura of slowing of motion so that he felt in full control of every ball that reached him. He too regretted the brevity of his experience.

B. Negative Experiences

The negative experience, occurring more frequently, was often clearly related to emotional conflict and stress, repeated and/or long standing.

Case 4, age 42 at time of surgery, had suffered a brain abscess at the age of 10 and had lost his mother during the convalescent period. Subsequently, he had a repeated experience of dread. Several nights a week, he would be left alone in the house while his grieving father went out to socialize until the early morning hours. Unable to sleep, the child would stand at the window, watching for his father to appear over the hill. To his mind came thoughts of violence that might have befallen his father and prevented his return. When his father would finally appear, the child, with great relief, would sink back in his bed and fall asleep. Early in adolescence, seizures appeared, manifested by an aura described as a feeling of "awesome terror," followed by a period of automatism which consisted of a pantomime appropriate to the feeling in the aura: He would grimace, shriek violently, breath rapidly, and pace up and down while beating his chest. Such behavior was frightening to all observers. He was amnesic for the behavioral sequence. Previous authors, such as Falret (1861), have noted that the seizure content may be related to dynamic events occurring at the time of initiation of the seizure disorder. Although it was known that this person had seizures, the behavior was interpreted by his psychiatrist as an expression of his relationship with his controlling wife. After eradication of seizures by surgery, the behavior disappeared completely (despite continuation of the marital relationship).

Case 5, was a patient with a lifelong problem in handling his ambivalence and aggressive feelings. He had developed a reaction formation of always being on good behavior, doing exhaustively for others, complying without questioning. In his automatic state, he would walk from room to room, striking anyone in his path.

Case 6, an adolescent with great family conflict and feelings of rejection, had an aura which produced significant suffering. He would develop fear and paranoid feelings—assured that anyone in sight was about to do away with him in a violent way. He stated that he would gladly lose a limb if doing so could substitute for this repeated dreadful experience.

There may be a clear short-circuiting effect with primary emotional thought emerging, as manifested by a sudden shift from rational to impulsive, often contradictory, verbal expression or behavior.

Case 7, Directly following a small seizure, the spontaneous recital of one patient shifted from an appropriate, reality-oriented evaluation of her situation to her usually unexpressed emotional feelings in that situation. Thus, at first, the patient spoke of her gratitude for a suitable job which had been secured for her with considerable effort. After a small seizure, as if unaware of her previously expressed attitude, she said vehemently that she would have to leave this office and voiced her criticisms, with paranoid ideation, about the extent to which she was singled out in subtle ways by her fellow employees.

Penfield (Penfield and Perot, 1963), early, spoke of the experiential seizure as an activation of old experiences. Later he said that in the seizure there might be "(. . .) alteration of present experience."

For example, in **Case 8**, a young patient, on her hesitant way to the dentist, became automatic and awakened with complete amnesia for the events, back at home in bed.

In **Case 9**, a young bank employee, extremely meticulous and correct to a compulsive degree, was sent to an important formal meeting in a distant city. As the meeting was prolonged, everyone no doubt felt restless. The patient became the unwitting spokesman, as he became automatic and suddenly arose, closed his briefcase and left, after saying loudly, "Let's get this thing over with."

C. The Destructive Nature of a So-Called "Minor Seizure"

Case 10, a young girl, while riding in an automobile with her parents, became involved in a rather heated discussion as to their subsequent direction. The patient had the feeling that she would like to get out of all this and, on a bridge, with the car proceeding at 50 miles an hour, she became automatic and was just saved from a serious accident as she attempted to open the door and escape.

Similar observations have been made by other authors: Ervin (Epstein and Ervin, 1956) reports the activities of a maiden lady who, during her seizure, would wander down the street to the home of a bachelor, whom she had been viewing with interest for many years, and undress in his front yard.

1. *Direct Expression of and Handling of Instinctual Needs Involving Hunger, Elimination, Sexual, and Aggressive Feelings as a Seizure Phenomenon or During Brain Stimulation*

Case 7, a 20-year-old girl, was accustomed to returning from her job at the same time as her working parents. The usual routine was for an obviously necessary cooperation in preparation and waiting for the meal. One evening, the family returned home, as usual. The patient stated that she was hungry, but contrary to her usual reaction, she became suddenly irrational, demanding food and, shortly, physically attacked her mother, whose explanations and entreaties were of no avail. As her behavior subsided, she took a short nap (probably postictal) and then joined the family in the delayed meal.

In regard to aggression, there is the question of differential diagnosis: Is the behavior due to emotional conflict only and therefore needing psychotherapy and/or medication, or is the negative behavior exhibited only when controls are diminished by epileptic discharge? Ervin and Mark (Mark and Ervin, 1970) at the Massachusetts General Hospital produced a book, "The Anatomy of Violence." They had studied patients with limbic system disorder manifested by psychomotor seizures and murderous impulses. Their subjects were criminals and, also, patients who came voluntarily because of fear of these impulses which might lead to acts of violence.

2. *Delayed Response to an Urge for Elimination by Hallucinatory Experience on Brain Stimulation*

Case 11: During corticography, a patient expressed a need to urinate and was told that a nurse would bring a bedpan. Meanwhile, another stimulus was applied to the amygdala. Suddenly, there was a change in her expression and conversational theme and she was obviously having a vivid visual hallucination as she said that she could see one of her parents emerging from the bathroom at home and walking down the hallway. A few moments later, when the bedpan appeared, the patient was surprised and said that there was no need for this.

D. Profile View of the Patient's Behavior over Periods of Weeks and Months

If one taps family reports and reactions at regular intervals, a fluctuating picture appears, as reflected in such remarks as, "If only he could be so cooperative and pleasant to live with all the time," alternating with, "You have to do something; he is impossible."

Case 12, a young man institutionalized at a state psychiatric hospital was seen at regular intervals in the seizure clinic. His family corroborated our observations of just such marked fluctuations in relation to frequency of small seizures. He would have days of very acceptable behavior alternating with days when he would threaten or become belligerent and attack patients and attendants, swearing and yelling until it was necessary to transfer him to a ward where there was greater security. Such disorganizing but subtle seizures have been called "minimal," "subclinical," and so on.

E. Alteration of Mood May Occur with the Sign Shifted After a Seizure

Case 13: When seen at varying times, this patient would be devastated by, or coping surprisingly well, with the same set of difficult circumstances, depending on the stage in his cyclical mood disorder.

Case 14: The seizures of another patient were anticipated by the family by 3–4 h during which he would become "either gay or ferocious." The mother stated that she would use all of her diplomatic powers to keep the patient from interacting negatively with his siblings until the mood and subsequent seizure had passed.

Some patients become depressed and long for a mood-elevating seizure: **Case 15**: Usually quiet, sensitive and introspective, and given to writing poetry, another patient would experience days of hyperactivity, when she would perform great feats of housework and philosophize by the hour. "At those times, I need a soapbox."

Case 16: Patient's response to an aphasic aura may be obstructive behavior, considered willful by the environment. For example, such behavior brought this adolescent patient to a psychiatric facility from which she was referred for investigation of seizures.

IV. The Interictal Period

There is considerable controversy in the literature regarding the so-called "interictal period." Behavioral disturbances have been explained by the seesaw theory (Landolt, 1953, 1958; Wolf, 1991) which claims that seizures alternate with disturbed behavior, that is, when seizures are under control, there may be behavioral symptoms, and vice versa. One would have to ask if seizures are truly absent or whether they are occurring in association with the disturbed behavior but are not recognized. In fact, major seizures may be absent while more subtle limbic seizures are occurring. Post (Post *et al.*, 1982) pointed out that major seizures can be anticonvulsant to limbic seizures. Thus, without major seizures, limbic seizures may be more frequent. The subtlety of limbic seizures has been described earlier. Terms such as "subclinical" or "larval seizures," "preictal or subictal excitation" attest to recognition of the difficulty of accurately ascertaining the occurrence of seizures. One may well question the designation "interictal."

A. Disturbances Occasioned by Anatomical or Functional Loss of Neurons May Result in Chronic Organic Mental Impairment of Varying Magnitude

The organic mental state acquires specificity from the involvement of higher cortical functions related to the temporal area of the brain including recent memory impairment, time sequence disturbance, auditory receptive dysphasia, and disturbance of critical comparative ability or ability to integrate complex experiences. These organic deficits may have corresponding behavioral ramification. The defect of recent memory leads to lack of continuity in experience. Time sequence disturbance may lead to disorientation or double orientation and a bias on association formation. Receptive dysphasia may produce comprehension difficulty, paraphasia, and loquaciousness. Disturbance of critical comparative ability may lead to fractionation of experience in terms of the individual's psychodynamic patterns, or response to single facets of qualities of objects in the environment. Ready incorporation of environmental stimuli occurs without recollection of the process. Subtle involvement of these higher functions may lead to manifestations of "everyday psychopathology" in these patients who may be described by family and peers as "different, queer, immature, using poor judgment, stubborn, disregarding logical arguments, slow to catch on."

B. Psychotic Features

Some psychiatrists have given such patients the label of schizophrenic. Slater *et al.* (1963) first described what they called "the schizophrenia-like psychosis of temporal lobe epilepsy." They differentiated this psychosis from classical schizophrenia by certain features, outstanding of which was the presence of organic mental findings. Other differences were absence of autistic features, maintenance of ability, and search for social contact. The authors would discard the schizophrenia label and speak of an "epileptic psychosis." The thesis that the psychotic picture arises from involvement of temporal lobe structures is strengthened by the fact that similar mental content often appears in the spontaneous seizure, in dream material, and in the response to electrical brain stimulation (EBS)—all experiential phenomena clearly arising from this region of the brain. In 1969, Dr. Rayport and I (Ferguson *et al.*, 1969), with other collaborators, reported on five cases whom we observed for some time prior to as well as during and following the development of a psychotic state.

V. Complete Case Histories of Two Patients with Psychotic Manifestations

A. Case 17

1. *Seizure History*

A 14-year-old patient was admitted to hospital for consideration of neurosurgical treatment for intractable temporal lobe seizures. Febrile convulsions had occurred from the age of 16 months to 3 years and psychomotor seizures had been present for 5 years (since age 9). The patient strongly demanded an operation in response to depression, saying, "Why can't I be like the others? I feel I am crazy," and reported violent thoughts and dreams which expressed her fears of serious injuries during a seizure, and so on. During ictal prodromes, which lasted for several hours, her personality would change from a child willing to cooperate and apologetic for being a burden to a child described as "unusually irritable, uncooperative and nasty." In her aura, she felt, "a funny feeling in my stomach, of seeing somebody dead, something disgusting, something red, I hear somebody say something." During a seizure, the family would observe simple automatisms (alimentary pantomime) or complex automatisms in which she would walk out of her way if she were alone, and, rarely, generalized convulsions. Following the automatism, the patient would complain of headache, would require several hours of sleep, and would be amnesic for the episode and for events preceding it by several hours. The frequency of seizures varied from an average of three to five per day during her menstrual period. Occasionally, she would have seizure-free periods of 1–2 weeks.

2. *Past History*

The patient's parents were separated. The father was institutionalized with the diagnosis of paranoid schizophrenia. The patient was living with her mother, a maternal aunt who was a mental retardate, and siblings. Her public school attendance in regular classes and a semblance of normal social activities were possible because of close supervision by the school staff and by two or three vigilant and devoted classmates. Two sexually slanted incidents were vividly remembered. When the patient was 4 years old, she had wandered into the room scantily dressed while the household was getting ready for a wedding. An adult visiting the family showed broad disapproval by gestures and with words of "shame, shame." When she was 8 years old, she was threatened sexually by her father. In her dreams, the patient would be killed by a truck. Since onset of the seizure disorder, her mother had warned her never to go out alone for fear of injury during a seizure.

3. *Treatment Course*

The patient underwent a resection of the right anterior temporal lobe, the amygdala, and the anterior 1.5 cm of the hippocampus. Postoperatively, the patient was found to have a memory disturbance characterized by marked impairment of retention of recent events and of sequential ordering in time. Not recalling the family's frequent visits, she would accuse them of not coming to see her. She was unable to correctly identify the hospital and most members of the staff. She did not remember that she had had an operation. She was terror stricken by and vehemently resisted routine postoperative procedures by doctors even though they were assisted by nurses. She screamed, "Don't do that to me." Lacking the recollection of her recent hospital course, she apparently misidentified the staff in terms of association of the childhood trauma to sexual threats. This psychiatric examiner, who came daily, was not recognized, being variously called "teacher," if she was carrying a portfolio, or "mother of another patient," if a visiting mother passed during the interview. She seemed deprived of reference to an established concept of self: When marked ataxia developed postoperatively from dilantin intoxication, she saw herself as "crippled," in contrast to a patient's usual detection of a pharmacogenic disability as being externally imposed and welcomed with the joy of a convalescent, but without insight, the progressive return of normal ability to walk after the intake of the drug had been lowered.

4. *Comments*

In this case, partial, unilateral hippocampal ablation unexpectedly produced the defect of recent memory described by Penfield and Milner (1958). The psychosis which developed after operation was noteworthy because of the interpretation of events in the current clinical environment in terms of remote amnestic material. Recent experiences not being "recorded" were unavailable for the formation of associations. The dynamic content of the available memories induced alarming results such as fear that the actions of the ward staff for a change of head dressing portended a sexual assault. At other, more neutral, times, people were identified by stimuli in the ongoing context; thus, the misidentifications reflected both prominent psychodynamic themes and the consequences of the deficit of memory for recent events. The association of the memory defect with impairment in the sequential ordering of events led to a loosening of ego boundaries. The patient's role-taking behavior no longer constituted an integrated type of response appropriate to the level of her personal development, but was directed by immediate clues in the internal environment, such as the toxic ataxia, or externally. For example, when a recreational therapy volunteer furnished her with coloring books and crayons, the well-meaning gesture conveyed to the patient that the brain operation had caused her to become mentally retarded, as feared and expressed preoperatively by her mother. During a 3-year follow-up, seizures recurred with lesser frequency and intensity. Psychiatric and memory difficulties gradually abated, but tended to recur in conjunction with increase in the seizure rate.

B. Case 18

1. *Seizure History*

A 31-year-old single man had had chronically recurring seizures since the age of 12. Attacks would begin with a visual aura of objects looking very far away, small, and green, followed by a sensation in his left hand: "Like a weight that shouldn't be there, as if my whole arm was bandaged." During the ictus, there would be elevation of his left arm, which he would grab with his right hand. Version of his head to the right would occur and, occasionally, progression to a generalized convulsion. The patient reported two dreams: In a recurrent childhood dream, which disappeared after the onset of the seizure disorder, there was involvement of limbs in the same manner as in the seizure. The patient would see himself going to a Chinese laundry and raising his arm to throw a rock into the shop. The laundry man would pick up an iron and raise his arm as if to hit back. The patient's associations to this dream consisted of a forbidden childhood activity of playing in a deserted field. The other dream was reported to one of us during a previous neurological hospitalization elsewhere. Past and ongoing emotional events were clearly presented. The manifest dream content was as follows: He was in a bar: "I asked for a drink—a 7 high. The man didn't know what I was talking about. He had on a white jacket. There was just the outline of his face and neck. Everybody was a stranger. I was all alone." The current negative association was his sensitivity to what he considered the dilatory response of the hospital staff to his request for attention. A positive and possibly dysphasic past element was the choice of the drink "7 high" which the patient had enjoyed in California during the most satisfactory phase of his life.

2. *Past History*

The patient completed trade school and spent 3 satisfactory years living with his brothers and working in their auto repair garage in California, until an increase in the frequency and severity of his seizures made even this sheltered employment hazardous for him. He returned to New York and became a recluse at home. Here he was overprotected, eventually becoming overly sensitive to any evidence of diminished attention, at times to a paranoid degree. Five years before admission to our hospital, he entered an epilepsy facility when the family could no longer provide adequate supervision. Episodes of marked impairment of memory had been noted by the family, by the institutional staff, and by us, when he came for regular clinic visits.

3. *Examination*

The patient was circumstantial, gave paraphasic responses related to impaired receptive ability. He showed impaired integration of complex experiences as indicated by misinterpretation of or failure to satisfactorily understand conversations, television programs, and items of formal testing. Intracerebral depth electrodes were implanted on the left into the amygdala region and anterior hippocampus. During a psychiatric interview, electrical stimulation of left amygdaloid contacts evoked an olfactory response described as "a smell of fish," followed by a vivid visual experience of a scene with "green water, people of small size fishing from a pier and a Howard Johnson restaurant." The micropsia and visualization of green resembled the components of his seizures. The mention of a Howard Johnson restaurant was related to the ongoing context. Earlier, in a psychiatric interview, the patient's reports of the olfactory sensations evoked by EBS had elicited from the psychiatrist a humorous comparison to the multiple ice cream flavors featured by the Howard Johnson restaurant chain. This comment had evoked strong and emotionally charged mirthful warmth among all the participants in the team interview situation. Some days later, while speaking of that experience, the patient reported a relevant and pleasant memory dating from his childhood, prior to the onset of his seizure disorder, when he had been able to go fishing without supervision at a dock near a Howard Johnson restaurant. Also, for several weeks after that brain stimulation interview, the patient reported hypnagogic experiences and, later, vivid daytime fantasies in which the same fishing scene would suddenly recur, but modified by confabulations related without the patient's conscious association to current situations on the ward. When he again reported the fishing scene, after several days of difficult interaction with a blond fellow patient, he stated that, "for some reason" the people in the familiar fantasy now had blond hair.

4. *Treatment Course*

In July 1964, left temporoparietal craniotomy was performed under local anesthesia. During surgical preoperative skin sterilization, a new agent had produced vigorous smarting of the freshly shaved scalp. The bright operating lights were trained on his head and face and the sheets, having slipped, he was marginally covered. The electrocorticogram revealed diffuse spiking so that no resection was carried out. During the procedure, he had a number of typical seizures associated with spike discharge and hyperemia of the face area.

5. *Early Postoperative Period*

The patient developed an amnestic psychosis with paranoid delusions and hallucinations which lasted for several weeks. He had no recollection of the previous 5 years, neither of his history preceding his admission nor of the operation. He felt that the hospital was against him because he had refused to accept a check given in payment for a picture of him that had been taken of him in the blazing sun when he was wearing a special suit with only an opening for his penis. He was sure that the blazing sunlight had caused a burn especially on his legs. He felt that the doctors were neglecting him because he was a Jew.

6. *Comments*

In this case, the psychotic reaction began following craniotomy during which there was a cluster of seizures. In the psychotic symptomatology, amnesia for recent events was associated with the symbolic appearance in delusions of recent and older emotionally charged but consciously forgotten material. In his delusional system, the operating room experience appeared in full array: The operating light became "bright sunlight"; stinging of the shaved scalp from the antiseptic became "a burning sensation," displaced to the genital area. Sensitivity about lack of attention to him, a theme which had also appeared in his dream content, was reflected in paranoid beliefs that everyone in the hospital was against him because he was a Jew. Spatial disorientation and a disturbance of time sequence were also manifested and interacted with the memory impairment. After a time, the patient believed that his religion was changed and that this conversion might propitiate the doctors who were against him. Conclusive evidence for the conversion was a rosary he had found on the bedside table which he believed to be his, but which, actually, belonged to the former patient in the now unoccupied neighboring bed.

Similarly, during EBS, his responses contained material from the interview itself, from older emotionally significant events such as the happy experience of fishing by himself before the onset of this seizure disorder and elements found in the aura of his seizures—that is, green water and micropsia. Later, the brain stimulation responses were elaborated by the unconscious incorporation of recent emotionally charged material into hypnogogic hallucinations and into active daytime fantasies. The recurrent dream preceding onset of the seizure disorder contained elements resembling the later aura of his seizures. [Epstein (1964) at Tulane University has written extensively to the point that recurrent dreams may be manifestations of temporal lobe disorder.]

VI. Conclusions

This chapter represents an interdisciplinary philosophy of approach to theories of brain organization. Clinical observations of patients with intractable temporal limbic seizures allow for the formation of hypotheses about brain–behavior relationships, particularly the structure of mind as theorized by Sigmund Freud.

History and observations of behavior, with attention to the timing of the behavioral symptomatology, provide insight into the interplay between unconscious, primary process thinking, and reality oriented responses, when the ego is in charge and defenses are fulfilling their protective role. Seizure discharge can be seen to release, probably by disconnection, expression of unconscious, primary process thinking, with emotional feeling tone consonant with the patient's true reactions, untempered by reality factors or by ego–superego shaping. Cognitive impairment, primarily of memory, contributes, at times, to the behavioral distortions.

References

Broca, P (1861). Perte de la parole, ramollissement chronique et destruction partielle du lobe antérieur gauche du cerveau. *Bull. Social Anthropol.* **2,** 235–238.

Doty, R. W. (1967). Limbic system. *In* "Comprehensive Textbook of Psychiatry" (A. M. Freedman and H. C. Kaplan, Eds.), Williams and Wilkins, Baltimore.

Epstein, A. W. (1964). Recurrent dreams; Their relationship to temporal lobe seizures. *Arch. Gen. Psychiatry* **10,** 25–30.

Epstein, A. W., and Ervin, F. R. (1956). Psychodynamic significance of seizure content in psychomotor epilepsy. *Psychosom. Med.* **8**(1), 126.

Falret, J. (1861). *In* 'De l'etat mental des epileptiques' Archives generales de medicine Ve serie Vol. 17, pp. 461–491. *In* "The Falling Sickness" (O. Temkin, Ed.), pp. 318. Johns Hopkins Press, Baltimore, February 22, 2006.

Ferguson, S. M., Rayport, M., Gardner, R., Kass, W., Weiner, H., and Reiser, M. F. (1969). Similarities in mental content of psychotic states, spontaneous seizures, dreams and responses to electrical brain stimulation in patients with temporal lobe epilepsy. *Psychosom. Med.* **31,** 479–498.

Freedman, A. M., Kaplan, H. I., and Sadock, B. J. (1976). "Modern Synopsis of Psychiatry." Waverly Press Inc., Baltimore.

Freud, S. (1953). Studies on hysteria. *In* "The Standard Edition of the Complete Psychological Works of Sigmund Freud" (J. Strachey, Ed.), Vol. 2, p. 608. Hogarth Press, London.

Freud, S. (1891). "On Aphasia: A Critical Study." German text translated and with an introduction by E. Stengel. Imago Publishing Co., London.

Gibbs, F. A. (1951). Ictal and non-ictal disorders in temporal lobe epilepsy. *J. Nerv. Ment. Dis.* **113,** 522.

Goldstein, K. (1995). "The Organism: A Holistic Approach to Biology Derived from Pathological Data in Man." Zone Books, New York.

Heilman, K. M., and Valenstein, E. (1993). Introduction. *In* "Clinical Neuropsychology" (K. M. Heilman and E. Valenstein, Eds.), p. 5. Oxford University Press, New York.

Jackson, J. H. (1875). 'On temporary mental disorders after epileptic paroxysms,' in West Riding Lunatic Asylum Medical Reports, Vol. 5, pp. 109–129. *In* "Selected Writings of John Hughlings

Jackson, Vol. I, On Epilepsy and Epileptiform Convulsions" (J. Taylor, Ed.), 1958, 119–134. Basic Books, New York.

Klüver, H., and Bucy, P. C. (1937). Psychic blindness and other symptoms following bilateral temporal lobectomy in rhesus monkeys. *Am. J. Physiol.* **119,** 3452–3453.

Klüver, H., and Bucy, P. C. (1939). Preliminary analysis of functions of temporal lobes in monkeys. *Arch. Neurol. Psychiatry* **42,** 979–1000.

Landolt, H. (1953). Some clinical electroencephalographical correlations in epileptic psychoses (twilight states). *Electroencephalogr. Clin. Neurophysiol.* **5,** 121.

Landolt, H. (1958). Serial electroencephalographic investigations during psychotic episodes in epileptic patients and during schizophrenic attacks. *In* "Lectures on Epilepsy" (A. M. Lorenz de Haas, Ed.), pp. 91–133. Elsevier, Amsterdam.

Luria, A. R. (1966). "Higher Cortical Functions in Man." Basic Books, New York.

Mark, V. H., and Ervin, F. R. (1970). "Violence and the Brain." Harper and Row, New York.

Moruzzi, G., and Magoun, H. W. (1949). Brain stem reticular formation and activation of the EEG. *Electroencephalogr. Clin. Neurophysiol.* **1,** 455–473.

Nielsen, J. M. (1936). "Agnosia, Apraxia, Aphasia: Their Value in Cerebral Localization." Hafner Publishing Company Inc., New York.

Ostow, M. (1954). Psychodynamic disturbances in patients with temporal lobe disorder. *J. Mt. Sinai Hosp. NY* **20,** 293.

Penfield, J. (1938). The cerebral cortex in man. I. The cerebral cortex and consciousness. *Arch. Neurol. Psychiatry* **40,** 417–442.

Penfield, W., and Jasper, H. (1954). "Epilepsy and the Functional Anatomy of the Human Brain." Little Brown & Co., Boston.

Penfield, W., and Milner, B. (1958). Memory deficit produced by bilateral lesions in the hippocampal zone. *Arch. Neurol. Psychiatry* **79**(5), 475–497.

Penfield, W., and Perot, F. (1963). The brain's record of auditory and visual experience: A final summary and discussion. *Brain* **86,** 595–696.

Post, R. M., Uhde, T. W., Ballenger, J. C., and Bunney, W. E., Jr. (1982). Carbamazepine, temporal lobe epilepsy, and manic-depressive illness. *Recent Adv. Biol. Psychiatry* **8,** 117–156.

Rice, J. L. (1985). *In* "Dostoevsky and the Healing Art: An Essay in Literary and Medical History," p. 84. Ardis, Ann Arbor.

Scoville, W. B., and Milner, B. L. (1957). Loss of recent memory after bilateral hippocampal lesions. *Arch. Neurol. Psychiatry* **20**(1), 11–21.

Slater, E., Beard, A. W., and Glithero, E. (1963). The Schizophrenia-like psychoses of epilepsy. *Br. J. Psychiatry* **109,** 95–150.

Weisenburg, T., and McBride, K. E. (1935). "Aphasia: A Clinical and Psychological Study." The Commonwealth Fund, New York.

Wolf, P. (1991). Acute behavioral symptomatology at disappearance of epileptiform EEG abnormality: Paradoxical or "forced" normalization. *In* "Neurobehavioral Problems in Epilepsy, Advances in Neurology" (D. B. Smith, D. M. Treiman, and M. R. Trimble, Eds.), Vol. 55. pp. 127–142. Raven Press, New York.

Zola-Morgan, S., Squire, L. R., and Amaral, D. G. (1986). Human amnesia and the medial temporal region: Enduring memory impairment following a bilateral lesion limited to field CA1 of the hippocampus. *J. Neurosci.* **6,** 2950–2967.

ID, EGO, AND TEMPORAL LOBE REVISITED

Shirley M. Ferguson* and Mark Rayport[†,‡]

*Department of Psychiatry, University of Toledo College of Medicine (formerly Medical College of Ohio), Toledo, Ohio 43614, USA

[†]Department of Neurological Surgery, Medical College of Ohio, Toledo, Ohio 43614, USA

[‡]Department of Radiology, Medical College of Ohio, Toledo, Ohio 43614, USA

This chapter offers further data supporting the hypothesis that there is a physiological basis for Freud's theories about the framework of mind. Data were gathered during our experience with patients treated by the interdisciplinary surgical epilepsy group at the Medical College of Ohio beginning in 1970.

In 1991, we presented a paper at An Historical Conference on Brain Functions Commemorating F. J. Gall (1791) and S. Freud (1891): "Is There a Neurophysiological Basis for Freud's Hypothesis of Ego Structure?" The following from the 1991 paper will provide the introduction to this chapter which expands on the hypothesis developed in the chapter "The Id, the ego, and the temporal lobe" by Ferguson and Rayport about the relevance of Freud's thinking for our observations in patients with temporolimbic epilepsy.

Dr. Rayport's introduction, here presented, sets the stage for the presentation of our clinical findings by reviewing Freud's biography as pertaining to the development of his "topographic theory" of the structure of mind.

I. Introduction

The objective of this presentation is to discuss the relevance of observations during electrical brain stimulation (EBS) in patients with intractable partial

INTERNATIONAL REVIEW OF NEUROBIOLOGY, VOL. 76
DOI: 10.1016/S0074-7742(06)76002-0

temporolimbic seizures to unconscious mental processes as these were elaborated and defined by Sigmund Freud. To accomplish this aim, we will briefly outline Freud's evolution from neurobiologist to neurologist to psychoanalytic psychiatrist. We will follow this overview with our recent rethinking of our epileptologic observations.

Dr. Ferguson and I have collaborated for three decades in the multidisciplinary study and treatment of persons with drug-resistant partial seizures. Acknowledgment is made of the participation of colleagues from two of our multidisciplinary teams, principally Dr. Robert Katzman, Dr. Morton Reiser, and Miss Molly T. Cox, REEGT at the Albert Einstein College of Medicine, and Dr. W. Stephen Corrie and Miss Carolyn A. Schell, RN at the Medical College of Ohio.

II. Historical Perspective

Sigmund Freud was born in 1856 in Freiburg, Moravia, and died in London in 1939 at age 83, having been ransomed in 1938 from the Nazi forces in Vienna through American effort.

Freud had enrolled in the Faculty of Medicine of the University of Vienna in 1873, where he was educated in the new intellectual and scientific climate produced by the biological investigations of Darwin and the studies in physiology and physics of the Helmholtz school which emphasized the unity of science, the operation of natural law, and scientific exactness. Theodor Meynert, Europe's leading cerebral anatomist, headed the psychiatry department. Ernst Bruecke, director of the Institute of Physiology, Meynert, and others with whom Freud came in contact believed that psychological phenomena could be explained in terms of physiological processes based themselves on chemical and physical laws. Freud received his medical degree in 1881 at the age of 25, having worked in Bruecke's Institute during 6 of his 8 years as a medical student.

In 1882, because of academic and personal circumstances, Freud embarked on a 3-year program of clinical training from which he emerged as a specialist in neurology. In 1885, Freud spent 4 months with Jean-Martin Charcot on the latter's service at the Salpêtrière, in Paris, while the latter was engaged in his ground-breaking studies of hysteria. Winson (1985) writes that, "Charcot provided a legitimacy for hysteria by demonstrating a rationale for its existence, a psychological rather than a neurological one." Under Charcot's influence, Freud developed a deep interest in the problem of hysteria and became firmly convinced that hysterical phenomena were genuine. Winson writes, "For the first time, Freud was exposed directly to the concept that (. . .) ideas of great potency (. . .) could exist in the mind without conscious awareness." Freud (1959) wrote in his autobiography: "I received the profoundest impression of the possibility that there could be powerful mental processes which nevertheless remained hidden from the consciousness of men." From 1887 to 1897, years during which Freud became aware of the cathartic method or "talking cure" used by his friend Joseph Breuer and began to systematically study the disturbances of his hysteric patients, psychoanalysis is said to have begun. Freud's significant and extensive written output of that period included the "Studies on hysteria" (Freud, [1895] 1953b), the "Project for a scientific psychology" (Freud, 1953a) written in two notebooks in the year 1895 and discovered among the papers of his friend Wilhelm Fliess 10 years after Freud's death, and "The Interpretation of Dreams" (1900) (Freud, 1953c).

Freud saw the study and interpretation of dreams in the context of the patient's associations as "the royal road to a knowledge of the unconscious activities of the mind" (Freud, 1953b). In "The Interpretation of Dreams," Freud set forth his "topographic theory" of the structure of mind. He divided mind into three regions: the unconscious, the preconscious, and the conscious. The unconscious contains repressed ideas ordinarily inaccessible to consciousness. These ideas can become conscious only through the preconscious, which excludes them by means of censorship or repression. Repressed ideas may reach consciousness when the censor is overpowered (as in psychoneurotic symptom formation) or when the censor relaxes (as in dream states). We submit, in addition, that the censor may be impaired by focal temporolimbic seizure discharge or a model of the partial seizure, discrete EBS resulting in brief after-discharge.

III. Clinical Material

Data are presented from encounters with 47 patients who had intractable temporolimbic seizures [one patient had bilateral frontal electroencephalographic foci and repeated episodes of status epilepticus (SE)] and who were referred to and studied by the members of the Comprehensive Surgical Epilepsy Program at the Medical College of Ohio between 1970 and 1993.

IV. Methodology

Every patient referred to the multidisciplinary treatment team underwent neuroclinical and neuropsychiatric interviews and examinations, interictal electroencephalogram (EEG) with extracranial electrodes, neuroradiologic and neuropsychologic studies. Beginning in the mid-1970s, the interictal EEG and chemical activation or deactivation procedures for localization of the epileptogenic focus were de-emphasized. They were replaced with long-term EEG and video recording of spontaneous seizures. In over 50% of our cases, localization of the epileptogenic focus required intracranial EEG recording from stereotactically implanted depth electrodes or subdural electrodes during spontaneous seizures. EBS was carried out in these patients to gain additional information about localization of the epileptogenic zone. Parameters for EBS were square wave pulses of 1 ms duration in 3-s trains at a frequency of 60 Hz. Constant voltage was increased in small increments until a cortical after-discharge was produced. To eliminate the possibility of kindling, after-discharges were elicited more than once only to resolve ambiguous ictal behavioral endpoints.

This methodology differs from that of other surgical epilepsy centers in one significant respect. During EBS, the patient is interviewed and observed by a neuropsychiatrist who has personally established a psychiatric database during extensive preoperative contacts, a unique approach, as far as we know.

V. Findings

Cases illustrate the various ways in which past or present experiences, thoughts, and feelings may appear during a seizure event. Emotional material which provokes the unique behavioral manifestations in each patient may have occurred repeatedly in the past, once in the past, or may be illustrative or be a response to current events in the patient's life.

A. Reflection of Early Trauma in Seizure-Related Behavior

1. *Case JAJ*

This 24-year-old man was considered for seizure surgery because of serious aggressive behavior which had caused him to be incarcerated in prison or in a psychiatric hospital 20 plus times since his teens. During an episode (ictal, postictal) he would display severe aggressive behavior toward people or things, whomever or whatever he encountered. It appeared as a release of tremendous, usually repressed, feelings of hurt and anger.

a. History. This patient was the 10th of 12 children. His seizures began in infancy. Since early life, he had experienced rejection by his family because of the burden of his epilepsy. Even when he was scheduled for brain surgery, the family refused to make contact. When he was helped to verbalize his feelings about his situation, he said he was regretful and hurt but had consciously accepted this as his lot. As in a subsequent case, he might also have felt guilt for the divorce of his parents which occurred when he was of age 11. His interictal behavior reflected his dilemma in his reaction formation. In the hospital setting, he was observed to engage in very pacific, feminized activity such as plant culture and needlepoint. He was soft spoken and diffident. In his interpersonal relationships, he was shy, but pleasant and warm.

b. Electrical Brain Stimulation. During stimulation of the amygdala there was a shift, from his usual pleasant affect and appearance to that of an individual with angry facies exhibiting a behavioral outburst. After spontaneous seizure events on the neurosurgical unit, he showed prolonged postictal irritability and dysphoria and only grudgingly, if at all, responded to attempts at conversation.

c. Outcome. During the first 2 years after left anterior temporal lobectomy, he was seizure-free; no adverse behavioral events occurred. Clearly, the ego was in command, as repressed emotional material no longer could break through.

2. *Case PAB*

This patient, a divorced woman of 27 years at the time of her right temporal lobectomy, came to our facility from another state where she had been in a psychiatric institution for 7 years because of her difficult behavior.

a. History. Her first seizure occurred at 16 months of age following head trauma. At 2 years, she was described as placing a hand on her upper abdomen, presumably indicating an epigastric sensation at the onset of her spells. At 10 years, irritability and rage outbursts appeared: "Sometimes, no one knew why she was mad." Early on, the family showed their impatience and rejecting attitude, which persisted into her adulthood and was observed by us. The seizures and behavior resulted in erratic school attendance. At 18 years of age, she was assigned to a sheltered workshop where she met a man in his 50s whom she married. He did not understand the illness. Her mother gave her a year's supply of antiseizure drugs as a wedding present. At 20, after bearing two children, this patient was admitted to a psychiatric hospital in her home state because of aggressiveness toward herself and others. She remained there for 4 years. At the age of 25, she was moved to our state psychiatric facility (through the help of a doctor who knew her and her family at the hospital and who had moved to Toledo) to be within the orbit of the epilepsy team at the Medical College of Ohio for possible surgical treatment.

b. Seizure Pattern. Her seizure pattern started with an aura of an epigastric sensation with fear, a paranoid feeling (that she was going to be attacked), visual illusions (walls were about to fall), and inability to speak. During the ictus, she would run, sometimes encountering dangerous situations. Postictally, she might attack other patients.

c. Observations in a Psychiatric Hospital. After a short period of time, the patient met with considerable rejection by patients and staff (as she had by her family and others in contact with her). She was considered unpredictable in mood, manipulative, not trustworthy, overly complaining, paranoid, and aggressive. Diagnostically she was regarded by some staff members as having a case of character disorder and/or psychosis as well as seizures: A "dual diagnosis." Observations on the psychiatry ward helped to identify an organic mechanism for some of these characteristics: (1) One day, the patient was found with bruises in an unexpected location on the hospital grounds. She stated that while out walking on the grounds, she had been attacked by a man and had been bruised in the struggle as she attempted to escape. Fortunately, a different, objective version was available. A nurse stated that she saw the patient walking on the grounds at some distance from a male patient. Suddenly, the patient began to run at top speed, tripped, picked herself up, and ran into the nearest building. This sequence was consistent with the description of her seizures. (2) She would get into frequent fights with other patients. Close observation showed that the individuals attacked were those with whom she had had previous negative interactions and this aggressive behavior occurred following a recognized seizure.

d. Observations on the Neurosurgical Unit. Marked fluctuations in mood state were observed: One day, she arrived in the EEG laboratory for EBS, morose, sullen, and irritable, exhibiting no spontaneous conversation. The next day, she was spontaneous, pleasant, and talking freely. She asked, "How is it you change? Yesterday, you looked sad. Today, you are happy. On some days, I feel on top of the world. On other days, I feel terrible, as if the world is against me."

e. Electrical Brain Stimulation. On stimulation of the left hippocampus, this patient had an epigastric sensation with fear. She stated that, "People look different." Looking at one of the medical staff, she said, "He is double his weight and strength," or "he is small. I could hold him in my hand. I hear people fussing. It is so fast, I can't keep up. The voices are tangled, as if someone is trying to overtalk you." On stimulation of the right hippocampus, she said, "Voices are screaming." This was followed almost immediately by her statement that, "the tone of voice is soft and pleasant." She then described a feeling of fear and said, "I'm afraid: I don't like to talk. I would like to get away. The words are twisted." On stimulation of the right amygdala, she became dysphoric and irritable: "I feel afraid. Someone is after me. I am going to be captured." The rapid fluctuations in affect and perception during EBS were similar to those appearing in her spontaneous behavior, which had resulted in a negative reaction from her environment.

f. Outcome. Following right temporal lobe excision (that included the amygdala and hippocampus) in 1975, her emotional lability and aggressivity were eliminated. It was possible for gradual rehabilitation to occur. She was helped to separate from the institutional setting, residing first in a group home, later at the YWCA, and, finally, in her own apartment. She progressed to a sheltered workshop and some competitive work. She expressed appreciation for her improvement in letters, poems, and during radio interviews, which became part of her active work in support of the cause of patients with epilepsy. She required no psychiatric treatment, but over the years made it clear (sometimes, with mild nonepileptic seizure manifestations) that she needed our continuing support. A dedicated counselor was of great help. Twenty years after surgery, she remarried. When her husband developed a kidney problem and required dialysis, she competently maintained his schedule of appointments, medications, and so on, provided a suitable diet, and appreciated the continuing companionship.

3. *Case BASA*

a. History. BASA, a 35-year-old married woman at the time of temporal lobectomy, for intractable seizures starting in childhood, knew rejection from early life. Her father, with whom she had had a mutually caring relationship, died during her childhood. Her mother, an unstable woman, did not hide her negative feelings about epileptics. She preferred her other healthy daughter and was hypercritical and clearly ashamed and rejecting of her epileptic daughter. The latter resented and tried to hide her illness. Nonetheless, she completed college, became an elementary school teacher, married, and had one child. However, scars were evident as she was shy in her interactions, had difficulty making eye contact, and restricted her interpersonal contacts. She also encountered difficulties in work.

b. Seizure Pattern. Her seizure pattern reflected a bimodal mood state. The prodrome consisted of depressive and paranoid components. The aura had successive dysphoric and pleasurable elements. Her seizures began with a feeling in her chest that "something bad is going to happen" and "a wave of strangeness." At times, this wavelike sensation would spread over her body, becoming pleasurable when it reached her genitalia. She noted that seizures were precipitated by excessive mood, positive or negative.

c. Electrical Brain Stimulation. After an increase in clinical and electrical seizures during intensive seizure monitoring with depth electrodes, the patient experienced a prolonged period (days) of hypomania characterized by increased psychomotor activity and elevation of mood with unusual giggling. Her speech was characterized by flight of ideas, clanging, paranoid ideation, and impaired social judgment. EBS of the left amygdala reproduced the affective content of her spontaneous seizures. Thus, increase in seizure frequency and EBS produced affective shifts characterized by alternation of depressive and pleasurable feelings. The psychosis, occurring after an increase in seizure activity, was an affective one of the hypomanic type lasting several days.

d. Outcome. After being rendered seizure-free, the patient became more confident, observed herself that she could make eye contact, and expanded her social sphere and was even able to motivate her diffident husband. When her mother became overtly paranoid and required institutional care, the patient experienced some guilt which she was able to work through in therapy.

She greatly appreciated the significant change in her life which successful surgery had made possible. No longer assaulted by fluctuations in mood, by poor self-image which had been reinforced by her critical, rejecting mother, by uncertainties about the response of others in view of the negative judgments of her mother, she was able to maintain a positive self-image, to relate more comfortably with others and over all, to function with greater success and satisfaction.

B. Seizure-Related Behavior Coincident with/Reflecting Current Emotional Stress

1. *Case MIR*

MIR, a 16-year-old boy at the time of study, was accustomed to frequent seizures, suffering from one to two minor seizures each night, two to three major seizures each week, and episodes of SE once a month.

a. History of Psychotic Episode. MIR had a history of seizures beginning at age 4. During his seizure he would have a prolonged automatism, during which he might wander outside, up to the altar of the church, and so on. On one occasion, he was brought to the Medial College of Ohio after one of his episodes of status had not responded as usual to treatment in a local hospital emergency room. The staff there had found that once control of the seizures had been achieved, his behavior was bizarre. Although fully oriented, calm, and able to recall recent events, he was markedly paranoid with persecutory and grandiose delusions. Psychotic themes were religious and sexual. He thought that the devil was trying to gain control and he was convinced that he was the transmitter of God's will. He had auditory and visual hallucinations with content related to the themes mentioned. He had illusions, thinking that there were cameras in various corners of the nursing unit taking pornographic pictures. His thinking normalized in about 5 days. Interviews with the family provided the current stressor which was affecting the whole family: The patient's sister had been involved in an illicit relationship and the fundamentalist church, in which the whole family was involved strongly, preached about the workings of the devil.

2. *Case HAJ*

This 41-year-old male farmer was referred for consideration of further surgical therapy for his seizures. At age 32, he underwent left temporal lobectomy at another institution and was seizure-free for 1 year when chronic seizures recurred. Resection included the amygdala, hippocampus, and a calcified hemangioma of the posterior fusiform gyrus.

a. History. His seizure disorder began at age 24 while he was in military service. The seizure pattern remained unchanged after the surgery.

b. Seizure Pattern. His seizures began with an aura of *jamais vu*, which was followed by a prolonged automatic state. He had had no psychiatric symptoms.

c. EEG Monitoring. Evaluations and observations during long-term scalp EEG monitoring revealed chronic impairment of recent memory (corroborated by neuropsychological testing). The patient had an increase of seizures, and, then, became disoriented and delusional. He declared himself to be the Lord. He left his bed periodically and strolled around the ward, confidently ministering to His "flock" of fellow patients whom he nonetheless greatly frightened. Attempts at restraint were resisted with blows. HAJ vigorously choked and slapped one of our colleagues and struck and stunned our nurse. He was transferred to a psychiatric facility where he had to be kept in physical restraints for 10 days. The patient was treated with anticonvulsant and antipsychotic medications. He recovered and was discharged after 3 weeks.

d. Electrical Brain Stimulation. One year later, persistent seizures led to the decision to proceed with stereotactic EEG to localize the intractable seizure mechanism. Prior to EBS, the patient was oriented and in a positive mood state. Following stimulation of the right hippocampus, he was asked about his head dressing. He said, "It's a bandage. I fell. I tried to get away but I ran into a two-by-four." After stimulation of the right lateral temporal cortex, he was asked where he was. He said, "In Hollywood. The police are here to make a film. I'm here to take photos because of blackouts."

e. Comments. In this case, EBS produced increased higher function deficits and paranoid thinking, antecedents to his previous psychotic episode. Even though EBS had been kept to a minimum, consistent with diagnostic objectives and avoidance of a possible prolonged psychotic episode, the disorientation and misinterpretation reappeared intermittently over a period of several hours.

It could also be seen that the context played a role in the psychosis described. The patient was a hard-working, freely moving farmer who had great difficulty in accepting the confinement of hospitalization. He had chronic memory impairment as well. His seizures were ushered in by an aura of *jamais vu*, followed by a prolonged automatism during which he would carry out his intent or feeling prior to the seizure. In the psychosis, he became "free"—he assumed a powerful stance and wandered masterfully and aggressively around the ward.

C. Some Patients Draw Unusual Conclusions from the Enigmatic Nature of the Seizure Experience and/or to Its Frequent Occurrence

1. *Case GAD*

GAD, a chronically paranoid 35-year-old woman with posttraumatic temporal lobe epilepsy, did not complete the preoperative evaluation for surgery. She interpreted her seizures as instances of care by her guardian spirit.

During the amnestic period of her seizure, she would carry out activities she had planned before the ictus. Thus, if hungry, she might continue with preparation of the meal during the seizure. On recovery, she would credit the preparation to her guardian spirit. If preparation for a bath coincided with her pleasant olfactory aura, she would presume that a fragrant bath had been drawn for her. The magical sense of grandiosity and relationship with an extraterrestrial agency became very real to her. She developed grandiose delusions that she had been selected to draw up plans for interplanetary space stations and would show us her sketches.

D. Emotion Attached to an Event or Person in the Past May Be Revived During a Seizure

1. *Case LAS*

LAS, in his lower 20s, would attack his brother during a partial complex seizure. In giving his history, he would recall that the onset of his seizures occurred after a period of hyperventilation, undertaken during a childhood game suggested by this sibling. The patient would add, "He's OK, and I'm stuck with seizures."

2. *Case DRO*

DRO was married and father of several children. During his seizure, he would aggress one, always the same, teenage child. The latter would attempt to escape as soon as he was aware that his father was beginning to have a seizure. The patient revealed that this child had been unwanted, but that over the years, he had accepted him.

VI. Conclusions

Exploration of the patient's experience of his seizures in relation to his psychodynamics provides an illustration of the physiological underpinnings of Freud's ego structure. Reflected in the seizure-related behavior may be emotional trauma of early life, negative feelings toward specific individuals because of past incidents or situations, extrapolation from the content of the seizure, and/or current emotional stress. Elimination of the seizures made it possible for the patient to respond to life situations, with reality oriented ego function and without the disruptive intrusion of id and superego.

Freud wrote of unconscious elements reflected in verbal expression or behavior, but often repressed by an evaluative superego and an ego responding to the reality context. Our findings/observations/hypotheses are consonant with ideas of Freud and with those of contemporary investigators: In "Freud's 'Project' Reassessed," Pribram and Gill (1976) state that, "Freud felt the requisite knowledge to provide the organic substratum for his metapsychological concepts was not yet available but would be at some time in the future."

References

Freud, S. (1953a). Project for a scientific psychology. *In* "The Standard Edition of the Complete Psychological Works of Sigmund Freud" (J. Strachey, Ed.), Vol. 1, p. 283. Hogarth Press, London.

Freud, S. (1953b). Studies on hysteria. *In* "The Standard Edition of the Complete Psychological Works of Sigmund Freud" (J. Strachey, Ed.), Vol. 2. Hogarth Press, London.

Freud, S. (1953c). The interpretation of dreams. *In* "The Standard Edition of the Complete Psychological Works of Sigmund Freud" (J. Strachey, Ed.), Vol. 4, p. 608. Hogarth Press, London.

Freud, S. (1959). An autobiographical study; inhibitions, symptoms and anxiety; the question of lay analysis; and other works. *In* "The Standard Edition of the Complete Psychological Works of Sigmund Freud" (J. Strachey, Ed.), Vol. 20. Hogarth Press, London.

Pribram, K. H., and Gill, M. M. (1976). "Freud's 'Project' Reassessed: Preface to Contemporary Cognitive Theory and Neuropsychology," p. 162. Basic Books, New York.

Winson, J. (1985). "Brain & Psyche: The Biology of the Unconscious," p. 72. Random House, New York.

SECTION II

STEREOTAXIC STUDIES

OLFACTORY GUSTATORY RESPONSES EVOKED BY ELECTRICAL STIMULATION OF AMYGDALAR REGION IN MAN ARE QUALITATIVELY MODIFIABLE BY INTERVIEW CONTENT: CASE REPORT AND REVIEW

Mark Rayport,[*,†] Sepehr Sani,[‡] and Shirley M. Ferguson[§]

[*]Department of Neurological Surgery, Medical College of Ohio, Toledo Ohio 43614, USA
[†]Department of Radiology, Medical College of Ohio, Toledo, Ohio 43614, USA
[‡]Department of Neurological Surgery, Rush Presbyterian-St. Luke's Medical Center Chicago, Illinois 60612, USA
[§]Department of Psychiatry, University of Toledo College of Medicine (formerly Medical College of Ohio), Toledo, Ohio 43614, USA

In 1999, Dr. Sani, then a student at the Medical College of Ohio (now, completing his neurosurgical residency at Rush Medical College, in Chicago) applied to do a project with Dr. Rayport during his senior year. This chapter is a result of that collaboration and is based on findings in a patient whom Dr. Rayport and Dr. Ferguson studied at the Albert Einstein College of Medicine.

Dr. Rayport prefaced the paper as follows: "We dedicate this paper to SIF and to other patients not deprived of life by their epilepsy who, during their determined struggle to shed its pervasive burdens, contributed otherwise inaccessible correlations between cerebral neurophysiology and behavior and unique momentum to the search for understanding of the human brain. This is their immortality."

I. Introduction

The classical investigations of the sensory responses evoked by electrical stimulation of the human primary sensory cortices have resulted in the conceptualization

DOI: 10.1016/S0074-7742(06)76003-2

0074-7742/06 $35.00

that the nature of the evoked responses is determined by the site of application of the stimulus. A similarly consistent correlation between stimulation site and response character has been reported with stimulation of the amygdaloid region, resulting in olfactory and gustatory responses which were "invariably disagreeable" in quality (Penfield and Jasper, 1954).

However, psychiatric and interdisciplinary neurosurgical and neuropsychiatric studies of patients with temporolimbic epilepsy had shown that environmental and intrapsychic variables became incorporated into the mental content of spontaneous seizures (Ervin *et al.*, 1955; Ferguson *et al.*, 1969), episodic psychosis of epilepsy (Ferguson and Rayport, 1988; Ferguson *et al.*, 1969; Rayport and Ferguson, 2001; Rayport *et al.*, 1986), dreams (Ervin *et al.*, 1955; Ferguson *et al.*, 1969), and responses evoked by electrical brain stimulation (EBS) (Ferguson *et al.*, 1969; Mahl *et al.*, 1964).

The present investigation addressed the question of whether olfactory or gustatory sensations evoked by EBS would be susceptible to qualitative alteration by modification of the affective tone of the subject's mental content by a structured interview. We report on a case in which we had the opportunity to test this hypothesis and which occasioned a critical review of the literature. Extensive computer-assisted and manual search of the literature did not disclose prior investigation of this question.

II. Case Report

SIF,[1] age 31 years, single, right handed, left hemisphere dominant for language, had developed a seizure disorder at age 12 after a fall from a tree causing no visible head injury. He completed trade school. He worked for 3 years in a brother's automobile service station until increase in the frequency and severity of the seizures made it necessary for him to return to his parental home where he became a recluse. Five years prior to referral he had to be institutionalized in a state hospital for epileptics. The patient's initial attacks began with a visual aura of objects looking "very far away," small and green, followed by a sensation in his left hand "like a weight that shouldn't be there, as if my whole arm was bandaged." Subsequently, the visual component became less prominent. His seizures began with a burning sensation in either hand and visualization of a green color, followed by unconsciousness and automatism. During the progressive ictus, version of the head to the right would be associated with elevation of his left arm which he might grab with his right hand, and occasionally a generalized tonic–clonic convulsion. Interictal scalp encephalograms (EEGs) indicated a left

[1]Additional information about case SIF has been published (Ferguson *et al.*, 1969). The letters "SIF" are a privacy code assigned to this patient; they are not his initials as implied by punctuation added by others during the process of production of Ferguson *et al.*

frontotemporal sharp wave focus. Activation of encephalography with intravenous pentylenetetrazol resulted in a typical seizure without adding to the EEG information.

After preoperative evaluation, a left parietotemporal craniotomy was performed under local anesthesia. Spontaneous partial seizures recorded during electrocorticography from the cortex of the inferior parietal lobule were accompanied by hyperemia of the left postcentral face area. Multicontact intracerebral electrodes were inserted into the amygdalar region and the anterior hippocampus through the cortex of the second temporal gyrus 3.5 and 4.5 cm from the tip of the middle fossa (Rasmussen, 1975). Subdural multicontact strip electrodes were placed over the postcentral gyrus and over the anterior and posterior second temporal convolutions (Ferguson *et al.*, 1969). A similar procedure was performed on the right side 1 week later.

Intracranial EEG recordings showed that the principal interictal epileptiform discharges occurred in the left postcentral region. Infrequent epileptiform abnormalities were noted from the right amygdala. The left amygdalar channels were devoid of epileptiform activity. Spontaneous seizures began in anteroinferior left parietal cortex and propagated to both temporal lobes. During diagnostic mapping of the intracranial electrode system by EBS, olfactory, gustatory, or combined olfactory-gustatory sensory experiences (OGEs) were evoked from the first two contacts of the left amygdalar electrode. When these evoked sensations were of sufficient intensity for confident perception, the patient described them as unpleasant. Occasionally, an OGE was associated with visual imagery.

The patient's postoperative recovery was uneventful in the neurosurgical domain. In the neuropsychiatric field, he developed EPE with an increase in the frequency of his partial seizures secondary to diminished oral antiseizure drug intake in the postoperative phase. He recovered fully from the EPE when drug blood levels were restored to therapeutic range. He was returned to the care of the residential institution.

III. Methods

Preoperative neurosurgical and neuropsychiatric assessment methods and the design and construction of the subdural electrodes have been described in earlier publications (Ferguson and Rayport, 1988; Ferguson *et al.*, 1969).

A. Intracranial Electroencephalography

The intracerebral multicontact depth electrodes had a flexible shaft of eight tightly twisted stainless steel wires (0.09 mm OD) quadruple-coated with Formvar®. Each wire ended in a 1 mm-bare-steel contact. The contacts were spaced at intervals of 5 mm beginning with the tip of the electrode shaft. The depth electrodes were inserted by hand through the middle temporal gyrus into the amygdale and anterior hippocampus (Rasmussen, 1975). Intracranial EEG was recorded on a 16-channel electroencephalograph (Model VI, Grass Instrument Company, Quincy, MA).

B. Electrical Brain Stimulation

Uni- or bidirectional bipolar square wave pulses (duration 1 ms) were delivered from the first and second contacts of the left amygdala depth electrode from a constant voltage stimulator with radiofrequency isolation unit (Model S-4, Grass Instrument Co., Quincy, MA) at frequencies ranging from 30 to 60 Hz in stimulus trains of 3-s duration. Stimulus current was monitored with an AC current probe (Model P6019, Tektronix, Inc., Beaverton, OR). Initial stimulation frequency during the psychiatric interview of 30 Hz was liminal to evoke olfactory and/or gustatory sensations that the patient could perceive confidently and describe. When stronger stimulation was needed in the latter parts of the interview to avert evoked sensations too faint for description, only the stimulus frequency was increased in order not to alter the magnitude of the voltage field.

C. Interview Setting

The structured interview was conducted in the monitoring room of the neurosurgical nursing unit. The patient was well acquainted with the neuropsychiatrist who had met with him in consultations at another academic medical center, and with the neurosurgeon who was his attending physician. An atmosphere of mutual trust and friendliness prevailed. Laughs were shared. The patient had as if "joined the epileptology team" in the shared effort against his seizures.

The patient sat in his bed. The neuropsychiatrist was seated at the left side of the bed and led the structured interview. The patient had been instructed to report to the neurosurgeon any evoked OGE or other sensation as soon as he experienced it, interrupting the ongoing interview if necessary. To do so, he needed to turn his head to his right toward the neurosurgeon who was observing the interview through the glass window of an observation room housing the 16-channel electroencephalograph and stimulator.

D. Interview Content

The patient's life experiences were known in considerable detail through prior neuropsychiatric contacts over a period of 5 years prior to his admission.

A structured interview was designed, based on the neuropsychiatric data, with alternation between topics known to be "pleasant" or "unpleasant" in six cycles of about 15-min duration. After four cycles, the patient initiated the choice of topic, transforming the interview into spontaneous monologue and dialogue. The 16 subcycles introduced by the patient were of variable duration. No attempt was made to override the patient's initiative. The reasons will be discussed later.

IV. Results

The total number of brain stimulations was 41, resulting in the report of 32 olfactory, gustatory, or combined olfactory-gustatory responses. Of the remaining nine stimulations, five produced other sensory responses and four produced no evoked response. Eighteen responses were olfactory. Twelve of these were adequately described for qualitative classification of "pleasant" or "unpleasant." Resulting from the use of liminal stimulation, six responses were described as "too faint" or "too brief" and were not classified. Of the 19 gustatory responses, only two were unclassifiable. All five of the combined olfactory-gustatory responses were adequately described.

Among the 24 adequately described olfactory and/or gustatory responses, 11 were "pleasant" and 13 were "unpleasant." "Pleasant" olfactory sensations were described as follows:

"Like a rose smell, . . ." (Stimulus 1)
"Smell all kinds of flowers. . .. I did see a yellow flower. It stood out amongst the others." (Stimulus 3)
"Smell, taste, and tingle of the tongue—sweet like corn—like a cornfield out in California." (Stimulus 16)

"Unpleasant" evoked sensations were expressed in the following language:

"Smell slightly, a bad smell. Out in California, had our bench with batteries to charge—it had the acid smell—that's how it smelled." (Stimulus 10)
"Taste and smell, like a flash, this may sound silly, but I did taste a flashlight battery." (Stimulus 31)
"Smell, like a woman who put too much powder on. Smelling the powder. Don't like it." (Stimulus 37)

Of the 12 evoked olfactory responses, 4 occurred during "pleasant" interview context. All were of "pleasant" quality, thus concordant with the context. None were "unpleasant" or discordant. Seven olfactory responses during "unpleasant" context were concordant, and one was discordant.

Of the seven gustatory responses, four were obtained during "pleasant" interview context. All four were of "pleasant" quality or concordant. Of three gustatory responses elicited during "unpleasant" interview context, three were qualitatively "unpleasant" or concordant.

Three combined gustatory-olfactory–evoked sensations were obtained in "pleasant" context and two combined responses in "unpleasant" context; each of these responses was qualitatively concordant with the interview context.

Among the 24 adequately described responses, 23 were of a quality concordant with the interview context and one was discordant. Of 11 olfactory and/or gustatory responses elicited during "pleasant" interview context, all were "pleasant" or concordant. Of 13 olfactory and/or gustatory responses evoked during "unpleasant" interview context, 12 responses were qualitatively concordant and only 1 was discordant.

V. Discussion

Smell and taste sensations preceding the onset of manifest seizures have been reported by epileptic patients predominantly as unpleasant, for many years. Hughlings Jackson recognized these conscious experiences as minor seizures. He adduced neuropathological evidence that these seizures resulted from lesional irritation of neuronal aggregates of the uncal region of the temporal lobe (uncinate fits) and contributed to the localization of primary receiving cortex for olfaction (Jackson, 1890). When evoked by seizure or electrical stimulation, these unimodal responses were devoid of affective components, except if pain was evoked as from the parietal cortex. At the modal level, the localization-driven paradigm appeared to fit evoked sensations in the viscerosensory modalities of olfaction and gestation.

The relatively greater incidence of unclassified olfactory responses as compared with the small group of undefined gustatory-evoked sensations may have been due to the use of liminal stimulus strengths, intrinsic response characteristics of the amygdaloid olfactory connections, or the greater environmental impingement on olfaction as compared with gestation. It is clear that if the six unclassified olfactory-evoked sensations were assigned to the discordant group, the qualitative concordance of olfactory responses with interview content would approach chance level. Contrariwise, if these unknown olfactory responses were concordant, the readiness of the olfactory system to qualitatively follow the interview context would be impressive. It is interesting that inadequately described gustatory-evoked sensations were few, eliminating this difficulty of interpretation with respect to the gustatory modality. All combined olfactory-gustatory–evoked sensations were few, eliminating this difficulty of interpretation with respect to the gustatory modality. All combined olfactory-gustatory responses were well described and tended to be associated with visual imagery.

A positive relationship between the interview context and the quality of the evoked olfactory sensations is nonetheless strongly suggested by the sequence of consecutive stimulation above liminal strength in which "pleasant" and "unpleasant" contexts alternated in the interview and were associated with a one-for-one alternation of qualitatively concordant responses. In contrast, it should be recalled that the apparently consistent "strong or unpleasant smells" of the literature were elicited by electrical stimulation during craniotomies under local analgesia. The comparison of the published data with the present data undermines the view that the "unpleasant" quality of evoked olfactory and gustatory sensations is an intrinsic functional characteristic of the stimulated neuronal aggregates and their connectivity.

We are unaware of similar observations in man on specific evoked sensations during monitoring of the antecedent programing and context of the sensory neural system stimulated.

VI. Summary and Conclusions

In a patient undergoing presurgical seizure monitoring for intractable partial seizures with implanted temporal and parietal depth electrodes, olfactory and gustatory sensations evoked by electrical stimulation of the amygdaloid region were modified by the affective tone of the subject's mental content in an interview setting addressing topics "pleasant" and "unpleasant" to him:

1. Both "pleasant" and "unpleasant" olfactory and gustatory sensations can be elicited by electrical stimulation of the amygdale.
2. At least for gustatory sensations, and probably also for the olfactory, the quality of the evoked sensation was modified by interview-induced changes in the affective tone of the patient's attention and mental content.
3. Prior descriptions of predominantly unpleasant sensations in man evoked by EBS and spontaneous partial seizures originating in the amygdale may have to be reconsidered in light of these findings.
4. Intensive computerized and manual search of the literature revealed no prior use of noninvasive modification of a subject's affective mental content to alter olfactory/gustatory experiential responses evoked by electrical stimulation of the nucleus amygdaloideus.

References

Ervin, F., Epstein, A. W., and King, H. E. (1955). Behavior of epileptic and nonepileptic patients with temporal spikes. *Arch. Neurol. Psychiatry* **74**(5), 488–497.

Ferguson, S. M., and Rayport, M. (1988). A multidimensional approach to the understanding and management of behavior disturbance in epilepsy. *In* "Modern Perspectives in Clinical Psychiatry" (J. G. Howells, Ed.), Vol. 15, pp. 302–330. Brunner/Mazel, New York.

Ferguson, S. M., Rayport, M., Gardner, R., Kass, W., and Weiner, H. (1969). Similarities in mental content of psychotic states, spontaneous seizures, dreams, and responses to electrical brain stimulation in patients with temporal lobe epilepsy. *Psychosom. Med.* **31**(6), 479–498.

Jackson, J. H. (1890). Case of tumor of the right tempro-sphenoidal lobe bearing on the localization of the senses of smell and on the interpretation of a particular variety of epilepsy. *Brain* **12,** 346.

Mahl, G. F., Rothenberg, A., Delgado, J. M., and Hamlin, H. (1964). Psychological responses in the human to intracerebral electrical stimulation. *Psychosom. Med.* **26,** 337–368.

Penfield, W., and Jasper, H. (1954). "Epilepsy and the Functional Anatomy of the Human Brain." Little Brown, Boston.

Rasmussen, T. (1975). Surgical treatment of patients with complex partial seizures. *Adv. Neurol.* **11,** 415–449.

Rayport, M., and Ferguson, S. M. (2001). Psychosis of epilepsy: An integrated approach. *In* "Psychiatric Issues in Epilepsy: A Practical Guide to Their Diagnosis and Management" (A. Ettinger and A. Kanner, Eds.), pp. 673–694. Williams & Wilkins, Baltimore.

Rayport, M., Ferguson, S. M., and Corrie, W. S. (1986). Contributions of cerebral depth recording and electrical stimulation to the clarification of seizure patterns and behavior disturbances in patients with temporal lobe epilepsy. *In* "The Limbic System: Functional Organization and Clinical Disorders" (B. K. Doane and K. E. Livingston, Eds.), pp. 171–182. Raven Press, New York.

SECTION III

CONTROVERSY IN DEFINITION OF BEHAVIORAL DISTURBANCE

PATHOGENESIS OF PSYCHOSIS IN EPILEPSY. THE "SEESAW" THEORY: MYTH OR REALITY?

Shirley M. Ferguson* and Mark Rayport[†,‡]

*Department of Psychiatry, University of Toledo College of Medicine (formerly Medical College of Ohio), Toledo, Ohio 43614, USA

[†]Department of Neurological Surgery, Medical College of Ohio, Toledo, Ohio 43614, USA

[‡]Department of Radiology, Medical College of Ohio, Toledo, Ohio 43614, USA

In 1992, this question of controversy in psychiatric diagnosis was explored at a Grand Round's presentation of the Department of Psychiatry at the Medical College of Ohio, Toledo, Ohio, by Dr. Ferguson.

This chapter provides an example of continuing controversy about the nature of behavioral symptomatology in the patient with temporal lobe epilepsy. Broad hypothesizing about behavior has often replaced observations made as closely as possible to the brain-mind interface. In the past, impedimenta are understandable since different types of epilepsy had not been defined nor was there any measure of brain electrical activity in order to correlate changes in brain function with behavior. In the present, behavioral data are often intentionally retrieved

INTERNATIONAL REVIEW OF NEUROBIOLOGY, VOL. 76
DOI: 10.1016/S0074-7742(06)76004-4

0074-7742/06 $35.00

as separately as possible from brain events with the rationalization that the behavioral observations not be contaminated by the effects of a seizure.

I. Introduction

The relationship between psychosis and epilepsy has been an enduring problem. It concerns such basic issues as the nature of the psychosis, mechanisms of its production, and its correlation with epileptic phenomenon. Is there a seesaw relationship between epilepsy and psychosis? Are epileptic patients psychotic as a result of seizure activity or are they alternately having seizures or psychosis because they are seizure-free? Is there an affinity or an antagonism?

In looking at the historical aspects of the relationship between epilepsy and psychosis, both possibilities are inferred. It must not be forgotten that different types of epilepsy had not been clearly defined nor was there any means of measuring brain electrical activity in order to correlate changes in brain function with behavior.

A relationship between epilepsy and psychiatric symptoms dates from earliest medical writings. Much of our information about the history of epilepsy in western civilization, dating from 400 B.C. to the nineteenth century, derives from the monograph entitled "The Falling Sickness" written by Temkin (1971).

Hippocrates (Temkin, 1971, pp. 4–5), contrary to the beliefs of his contemporaries, taught that the brain was the seat of all human experience, positive and negative, including the so-called "sacred disease" or epilepsy, and madness. Other early commentators noted the coincidence of epilepsy and insanity: Herodotus (Temkin, 1971, pp. 4–5) wrote of the mad behavior of a Persian king. "(. . .) from birth he suffered from 'a certain great disease which some people call sacred. And thus it would be not unlikely that if the body suffered from a great disease, the mind was not sound either.' "

To assuage the feelings of those so unfortunate as to be afflicted with epilepsy, modern writers have often referred to historical leaders, writers, and artists who had the disease. Among them were Caesar, Napoleon, Dostoevski, Van Gogh. There were religious figures as well. Thus, Mohammed is said to have attacks in which "It was said that his face took on a dusky, pale hue, that for an hour he resembled a drunken person, that sweat ran down him like pearls." And, "I saw a vision of an angel called Gabriel and not being able to bear the sight of him, I become weak and fall down" (Temkin, 1971, pp. 152–153). Dostoevski writes of the "(. . .) happiness which we epileptics experience for a second before an attack. Mohammed avows in his Koran that he saw Paradise and was in it" (Rice, 1985). These are examples of the recognition of religiosity in the thinking of epileptic patients.

We may now quickly pass over many intervening centuries to the nineteenth, when the history of modern psychiatry began. When Pinel in France and Tuke in England converted prisoners to patients, epileptics were among them. The epileptic was seen as having insanity complicated by epilepsy. The epileptics were separated from the other insane patients because they were found to be incurable and because it was considered an additional burden for nonepileptic individuals to view those seizures.

Hughlings Jackson (Jackson, 1875, p. 122), the great English neurologist and most important nineteenth century writer about epilepsy, made some very shrewd observations which, as not infrequently, return even now as newly discovered findings and hypotheses. He classified the association between mental states and epilepsy in terms of an affinity:

1. "The sudden and transient mental disorder after one or a few fits"
2. "More lengthy infirmity after the rapid succession of numerous fits"
3. "Persistent deterioration, (. . .) the result of fits repeated for months or years"

In regard to mechanisms of the psychosis he said, "I think it probable that there is a transitory epileptic paroxysm in every case of mental automatism occurring in epileptics before their mental automatism sets in" (Kraeplin, 1923).

The term epileptic equivalent was introduced by Hoffman (Temkin, 1971, p. 361) in 1862. He stated that "the epileptic attacks can be lacking altogether and yet by means of its peculiar symptoms the insanity may reveal itself as an equivalent of epilepsy, (. . .)."

Maudsley (Toone, 1981), for whom the great psychiatric hospital of London was named, well described the natural course of the untreated epileptic and emphasized the toll on mental function that may occur: "Gentle, amiable, and industrious through their long intervals of lucidity, they hope against hope that each recurring paroxysm will be the last; they eagerly try all remedies, in the hope of curing the disease (. . .) but confidence wanes as the attacks recur, the mind is slowly weakened by the storms of fury through which it passes, (. . .)." Although our medications are very effective in some cases, there are patients who continue to have seizures and may also be faced with behavioral accompaniments and repercussions such as psychosis.

Kraeplin (1856–1925; Landolt, 1958), the great psychiatric classifier, also studied epileptic psychosis. He declared it a major form of mental illness, alongside dementia praecox and manic depressive illness. He noted acute mood swings as well as hallucinations and delusions.

With all this, the burning question remained as to the precise relationship between the two entities with suggestions that psychosis might occur in association with clear epileptic activity and/or that seizures and psychosis might alternate.

Answers seemed in sight with the appearance of the modern era of epileptology which dates from the discovery of the human electroencephalogram (EEG) in 1929. Gibbs (Glaser, 1964), a neurologist/electroencephalographer working in Chicago, who defined the pattern of discharge in patients with seizures arising in the temporal lobe of the brain, noted the frequent occurrence of behavioral symptomatology in these patients. These were shown to have behavioral components, sometimes of a psychotic flavor. Gibbs (1951) wrote that temporal lobe epilepsy and psychosis are "commonly associated."

Subsequently, although neurological and behavioral disciplines became involved in the study of these patients, intellectual paths joined infrequently. Assessment of seizure activity was not often accompanied by rigorous clinical behavioral descriptions. The dearth of sophisticated collaborative work is reflected in the lack of description of the relationships of psychopathological and neurophysiological phenomena. Recently, in the climate of psychobiological understanding, the pendulum has swung in psychiatry to an interest and in fact a demand for precise behavioral description of symptoms. Revisions of the Diagnostic Statistical Manual (DSM) represent the culmination of this trend. Similar precision has yet to be extended to the definition of psychosis in epilepsy.

In epileptic patients, "psychosis" has been variously applied, sometimes referring to a distinct entity, epileptic psychosis, often episodic, but at other times to chronic psychosis such as schizophrenia and affective conditions. Patients with temporal lobe seizures often did not respond to medication and were referred for surgery.

In the 1950s, several English psychiatrists, all working with patients of the Maudsley Hospital, studied the psychiatry of temporal lobe epilepsy. Hill (1953) described a "paranoid hallucinatory psychosis." Pond (1957) emphasized the absence of affective impairment as in true schizophrenia.

Slater *et al.* (1963), also working in London, studied 69 patients. He introduced the term "schizophrenia-like psychosis." Symptomatology included prominent paranoid ideation, often with mystical and religious coloring. These states differed from classical schizophrenia in the preservation of affect, the presence of organic findings, and the absence of genetic tainting as well as of a premorbid schizoid personality.

His categories were rather impure:

1. Patients had repeated short-lived confusional episodes and then became chronically psychotic. This occurred in 11 patients.
2. Those who resembled paranoid schizophrenics: 46.
3. Those who resembled hebephrenic schizophrenics: 8.
4. Patients with petit mal epilepsy.

Symonds (1962), an English psychiatrist, in critiquing Slater's work, noted that the first three groups included patterns of psychosis, whereas the fourth, a type of epilepsy.

Bruens (1974) who wrote a definitive chapter on psychosis in epilepsy in the "Dutch Handbook of Neurology" reviewed Slater's cases and noted that primary delusions, a *sine qua non* for the diagnosis of paranoid schizophrenia did not occur and, in fact, that the delusions were "empathizable." He concluded that none of the psychoses fulfilled the criteria for the diagnosis of schizophrenia. Scandinavian investigators (Parnas and Korsgaard, 1982) described psychosis in epilepsy as the "paranoid hallucinatory syndrome of epilepsy." Despite all this, Slater's paper was very influential and his schizophrenia-like psychosis of epilepsy became often synonymous with psychosis in epilepsy.

In 1980, Perez and Trimble (1980) at Queens Square in London sought to remedy what they saw as a lack of accurate phenomenological description of the psychotic features by using a quantitative, computerized approach to diagnosis. In 24 consecutively referred cases with epilepsy and active psychosis, which had continued for more than 1 month, they found that the psychosis of 11 cases fulfilled the criteria of the nuclear schizophrenia syndrome. However, there was a comment that illustrates what we have said about the lack of equally careful diagnosis in both the psychiatric and epileptologic spheres: in describing the epileptic component, Perez and Trimble stated that these patients had "complex partial seizures of *probable* temporal lobe origin."

II. Classifications of Psychosis in Epilepsy

What are the psychoses of epilepsy? Classifications of the last two decades reflect evolution of our understanding.

Bruens (1974), in 1971, presented a classification based on the *state of consciousness and duration of the psychosis* as follows:

1. Psychotic states with disturbance of consciousness (twilight states)
   a. Psychoses in which the disturbance of consciousness is the most prominent feature
      i. postparoxysmal twilight states
      ii. absence status (Lennox), status psychomotoricus
      iii. twilight states not chronologically related to the seizures
   b. Delirious states
2. Psychotic states with normal level of consciousness
   a. Brief, often periodic psychoses (episodic psychoses)
      i. dysphoric states
      ii. manic states
      iii. depressive statesiv. alternating psychoses with productive symptomatology

b. More prolonged or chronic psychoses
   i. paranoid syndromes
   ii. psychoses characterized by marked psychical regression
   iii. schizophrenia-like syndromes
   iv. manic-depressive psychoses

In 1976, Betts (1981) proposed a classification based on *psychopathology*:

1. Global disruption of personality. There is disturbance of thinking, behavior, emotional relationships, feelings, and perception. This group includes the schizophrenias and schizophrenia-like states.
2. Primary disturbance of mood. There may be depression or pathological elation and such secondary symptoms as depressive hallucinations or delusions.
3. Organic psychosyndromes. These include confusional states, changes in behavior, in the context of epilepsy associated with subictal epileptic activity, and delirium.

In 1981, Toone (1981) offered a classification in which the psychotic states were differentiated by the *presence or absence of seizure activity*:

1. Psychoses directly related to the occurrence of seizure activity
   a. Abnormal mental states that are a direct and immediate consequence of underlying seizure activity and which may have some of the features of a psychotic state.
      i. continuous auras
      ii. petit mal status
      iii. temporal lobe status
   b. Postictal psychoses
2. Interictal psychoses
   a. Schizophrenia-like and paranoid psychoses
   b. Affective psychoses

That epileptic seizures can be associated with or cause psychosis is widely accepted when the immediate link can be observed. The challenge has rested with the determination of the pathogenesis of the psychotic states that have been described as chronic and/or interictal. Critical reservations with regard to the designation interictal have remained. In 1964, for instance, Glaser (1964) stated: "Often, however, it is difficult to separate an ictal or postictal psychotic episode from an interictal behavioral disorder." Acute epileptic psychotic reactions have been recognized as actual components of what is now regarded as the psychomotor-temporal lobe seizure complex since their first description by the French neuropsychiatrists of the mid-nineteenth century and the later analyses by Jackson, Penfield and Jasper, Gasteau and Lennox.

In 1969, we (Ferguson *et al.*, 1969) emphasized the role of subtle seizure activity in the production of psychosis.

III. Relationships Between Psychosis and Seizure Disorder

Hypotheses in the literature regarding the relationships between psychosis and seizure disorder have ranged from age-related coincidence (Stevens, 1966) to specific mechanisms which shall now be considered.

A. Pathophysiologic Mechanisms

Pond (1957) postulated that abnormal affect and/or perception might become integrated into the psychic life and become the anlage of psychosis.

Slater *et al.* (1963) made a jump from anatomical to mental change. Thus, he attributed psychosis to the dying cells of an atrophic process.

Symonds (1962) contradicted Slater saying that psychosis is caused not by loss of neurons but by disorderly activity of those that remain.

Flor-Henry (1969) was the first to show that epileptic activity in the dominant temporal lobe is highly correlated with psychosis of a schizophrenic type. There was both agreement and disagreement.

Toone (1981) tried to substantiate this finding by radiologic evaluations but found that there was no firm support for this hypothesis. It is to be noted that computerized tomography studies of schizophrenics without epilepsy confirmed the lateralization hypothesis but found localization of the pathology in the frontal lobe.

B. EEG and Clinical Seizure Activity

Landolt (1958) did serial scalp EEGs in 107 cases of epileptic twilight states and psychotic episodes. These episodes were characterized as "epileptic mania." There was an inverse relationship between seizure occurrence and the affective psychotic episode during which he observed forced normalization in the EEG.

One may refer to the long history of an antithetical relationship between psychosis and seizures as evidenced in the history of the development of electroconvulsive therapy (ECT). Whereas Meduna promoted camphor-induced seizures for the treatment of schizophrenia, it was soon observed that affective symptoms were more responsive and that only the catatonic symptoms of schizophrenia responded. The most effective response to ECT, even up to this day, is in patients with delusional depression. Post *et al.* (1982) suggested that major motor seizures, such as those artificially induced by ECT, might be anticonvulsant to seizure activity in the limbic system. It would not be contradictory that major seizures and antiseizure drugs (ASDs), such as carbamazepine, might be helpful for the control of affective disorder. If major seizures are controlled, disturbed behavior may occur due to the persistence of minor limbic seizure activity. This may, in part, explain the controversy over the antagonism or affinity of psychosis and epilepsy. It may also be noted that Slater *et al.* (1963) reported a decrease in major, not psychomotor, seizures in some epileptic patients who became psychotic.

C. Psychogenic Factors

Intrusion of seizure experiences, suggested by Pond (1957), may be compounded by chronic impairment of temporal lobe function leading to reduced adaptability and psychosis.

D. Higher Function Impairment

Several authors (Bruens, 1974; Ferguson *et al.*, 1969; Glaser, 1964) observed impairment in cognitive processes in relation to psychosis.

E. Pharmacologic Factors

Reynolds (1968; Toone, 1981) suggested that folate deficiency secondary to phenytoin might precipitate psychosis.

Trimble (1977) and Trimble and Meldrum (1979) proposed that a central increase in dopamine might raise the threshold for seizures and increase the propensity for psychosis.

These suggestions make possible the occurrence of psychosis in response to the treatment of the seizures, not the seizures per se.

In focusing on these findings, one must return to the clinical level and ask how is it that with similar pathology and biochemical change more patients with seizure disorder, controlled and uncontrolled, do not have psychosis?

F. Multifactorial Hypothesis

In contrast to a parsimonious explanation involving an anatomical structure or pathological change, neurochemical factor, and so on, other authors, including the present authors, have proposed a multifactorial hypothesis to account for the occurrence of psychosis in epilepsy.

In 1969, Ferguson *et al.* (1969) stressed the need for an individual approach leading to a multifactorial basis for epileptic psychosis, including seizure phenomena, cognitive change, and dynamic issues.

Bruens (1974) stated that, "A multifactorial genesis has to be considered, the various factors being of differing significance in each individual case." Fenton (1978) concluded that, "Like most psychiatric disorders, it is probable that the psychoses of epilepsy result from a complex interaction and multiple etiologic factors—biologic, psychologic, and social. The relative importance of any one factor varies from individual to individual with interplay of pathogenic factors."

Considering these differing views, the prevalence of interictal psychosis faces substantial problems. One consensus that has been reached has been that prolonged psychoses in patients with epilepsy occur predominantly in patients with temporal lobe epilepsy. The difficulty of accurate estimation of prevalence relates partly to the phenomenological problems discussed. Complicating the issue further is that previous figures largely represent the schizophrenia-like and paranoid psychoses often accepted as the interictal psychoses of temporal lobe epilepsy. Trimble and Perez (1982) in later work cautioned against this practice in their description of the psychoses of temporal lobe epilepsy as heterogeneous because of the presence also of affective disorders. In view of all of the obstructions and uncertainties, Fenton (1978) concluded that, "Little is known about the prevalence of psychosis in epilepsy."

IV. Experience of the Epilepsy Comprehensive Program at the Medical College of Ohio

One case from the Einstein series is included.

A. Philosophy

The authors' studies strongly support an affinity theory. We value our philosophy, expressed from the beginning by the Director of the program, Dr. Mark Rayport, which emphasizes the need for the interdisciplinary, thorough study of the individual patient. Highlighting the problems that have arisen in understanding this entity, we would point out that some authors writing about psychiatric symptoms in epilepsy have taken patients with undefined or unclassified seizures. On the other hand, the neurologist/neurosurgeon/electroencephalographer who analyzes the seizure pattern with great sophistication may be quite uncritical in the behavioral sphere. For example, in the surgical situation there may be no psychiatric baseline prior to operation, or, in clinical research studies, routine retrospective reports in hospital charts are accepted as psychiatric data.

B. Methodology

Patients with intractable temporolimbic seizures were studied intensively preoperatively by a team composed of a neurologist, neurosurgeon, neuropsychiatrist, neuropsychologist, clinical nurse specialist, and counselor. Studies continued for prolonged periods giving an opportunity to study behavior, including episodes of psychosis. In surgical patients, such observations were resumed postoperatively to assess the fate of any preoperative symptoms and/or the possible development of psychosis for the first time. This orientation has provided the opportunity to discern links between brain activity and behavior.

Postulating the intervening variables between the seizure disorder and psychosis is made possible by studying patients over time and in the framework of simultaneous observation of behavior and electrical activity recorded from the limbic areas and from the cortical surface. We have come to believe that the mechanisms may vary, often depending on the anatomical location of the seizure focus, the parameters of the electrical seizure activity, the experiential content of the seizure, frequency of seizures, cognitive dysfunction, and the intersect with the patient's life. We would designate as improbable a definite antagonism or seesaw relationship between seizure occurrence and psychosis. Rather, seizure activity and/or its repercussions may be unrecognized; patients declared psychotic and thought to be seizure-free may actually be experiencing seizure activity, perhaps of a subtle type. It should be pointed out that a fertile ground for the development of psychosis is essential.

V. Illustrative Case Material

A. Role of Psychosocial Factors

The following two cases illustrate the fact that even in the ictus-related psychosis, psychosocial factors may be essential for the precipitation of acute psychosis, once disruption by seizure discharge has occurred. The similarity between seizure phenomena and psychotic phenomena is evident. The role of religiosity should also be noted. Religious experiences were present in the reality situation and in conscious seizure experiences; they were incorporated into the psychosis and later into interictal thinking.

1. *Case MIR*

As described in chapter "Id, ego, and temporal lobe revisited" by Ferguson and Rayport, MIR was a 15-year-old male who had a period of psychosis [after one of many episodes of status epilepticus (SE)], which occurred in the context of an exceedingly stressful, emotional situation.

Recovery of insight was only partial as psychotic material lingered in his mind. He was not convinced as to the unreality of the psychotic events, "but felt that God had resumed proper control." During the subsequent several years, he continued to have episodes of SE about once a month, but did not become psychotic.

2. *Case RUD*

RUD, a 23-year-old man, began to have posttraumatic seizures at age 10. His seizures were introduced by an aura described as an epigastric sensation with fear, associated "with a feeling of guilt and a fear of dying." This aura articulated with a traumatic life experience consisting of parental strife that led to divorce despite strong Catholic affiliation. The patient felt guilty, thinking that he was responsible for his seizures and for the marital discord. Early, he sought solace in religion, trusting that the Lord would prevent the disruption at home. As official changes in church ritual appeared, RUD felt abandoned and was in considerable conflict over his religious practice. He finally resolved this by leaving the church.

During his psychosis, which followed a period of SE, he became very vocal, declaring that he was threatened by hell fires around him and warned the staff, whom he clearly recognized, to keep their distance for their own safety. He felt that Satan was trying to get control of him. He experienced various illusions that his hands were melting, his feet were like rock, and sounds were "weird." Upon recovery after 5 days, he recognized the abnormality of his thinking, but felt the experience could serve as a reminder that he should return to the church. He said, "I must get my act together." For a time after his psychotic episode, he read the Bible assiduously and attended church services regularly.

B. Role of Higher Function Deficits

1. *Case 11*

Case 11, mentioned in chapter "The id, the ego, and the temporal lobe" by Ferguson and Rayport, illustrates the misapplication of a schizophrenia diagnosis for a chronic organic behavioral syndrome and for acute episodes that could be shown to be clearly related to subtle ictal discharge. Such a sequence is not uncommon if a patient is treated on a general psychiatry service. In 1969, Ferguson *et al.* (1969) reported this case of a male patient who had had a seizure disorder from childhood. He was viewed by his desperate family as nonfunctional and unable to adequately process everyday situations. He was considered by some as chronically psychotic.

a. Seizure Pattern. Onset of the seizure was characterized by the sudden inability to understand what was being said or what he read. Later, he would say, "A line keeps coming back; I try to figure it out. I write it down but it doesn't make any sense." After a prolonged behavioral automatism, difficulty in remembering ongoing material would continue for several hours.

This seizure pattern indicated a dominant temporal lobe focus involving the posterior speech area as well as affective and integrating processes.

b. Psychosis. On two occasions prior to coming under our observation, this patient had been reported to have suffered a "schizophrenic psychosis" following a period of SE. We were able to observe the onset of a psychotic episode in the wake of an increase in minor seizures noted only by the more experienced staff members and in retrospect. The first behavioral symptomatology consisted of his eating hallucinated food particles after he had refused food offered to him by a nurse. Later, this behavior was interpreted as misidentification of the nurse as his mother with whom food had become a major area in the battle for control. When his behavior became unmanageable because of overactive destructive movements, he was sent to an isolation room in a general psychiatry ward where a diagnosis of schizophrenia, catatonic type, was made. In the psychiatric records, seizures were said to be absent.

However, when the epilepsy group studied the patient extensively on the psychiatry unit, he was observed, clearly, to be having frequent small subtle seizures. His condition was reinterpreted as an acute organic psychosis superimposed on a chronic organic brain syndrome. During gradual recovery over weeks, a resurgence of psychotic thinking was noted whenever small seizures recurred.

After he recovered, the patient reported that during the period of isolation, necessitated by his hyperactive destructive state, he had concluded (in concrete fashion) that he had misbehaved: "I was separated from my family to keep me out of the way." Items in the environment had been incorporated into his disordered thinking. A travel poster on the wall in the unit made him feel that he was a prisoner in a concentration camp in Europe.

c. Outcome. After temporal lobe excision, the patient became seizure-free. He no longer had the prolonged periods of memory difficulty and the aphasic experiences or psychotic episodes. In time, rehabilitation was successful and he was able to become independent and to work as an employee of a bank.

2. *Comments*

Higher function deficits associated with seizures and occurring in a person with chronic cognitive compromise and organic personality difficulty may introduce the aberrant thinking and loss of reality contact which may be interpreted by the nonepileptically oriented psychiatrist as schizophrenia separate from the epilepsy.

In such patients as the above, chronic psychosis was not a separate interictal event alternating with seizure occurrence. Rather, the psychosis was clearly related to seizure events, reactions of the environment, and cognitive dysfunction. Chronic psychosis was characterized by varying combinations of paranoid and obsessional thinking, delusions and hallucinations, somatic preoccupation, religiosity, magical thinking, and aggressive behavior. Subtle deficits of higher cerebral functions were frequent, including impairment of auditory comprehension with easy misinterpretation, compromised memory function, concreteness, and difficulty in getting the point. The intrusion of unusual ictal experiences was also evident in the psychotic state.

C. Intrusion of Unusual Ictal Experiences

1. *Case GAD*

GAD developed a psychosis after experiencing frequently occurring enigmatic seizures. (Detailed case description in chapter "Id, ego, and temporal lobe revisited" by Ferguson and Rayport.)

2. *Case WIT*

WIT, a 16-year-old girl, had had a behavioral problem since the age of 3 when she became recurrently aggressive toward other children. For this, she would be appropriately punished. At 5 years, she had her first recognized psychomotor seizure. During her aura, and also in her dreams, she would feel very afraid, as if being chased. In her general demeanor, she was chronically paranoid: She felt that she was being checked and picked on and was hypercritical of others. She related poorly to people of any age. Examination of higher nervous functions showed a subtle impairment of comprehension and of sequential ordering.

Early onset of frightening experiences, in both seizure phenomena and environmental response, seemed to initiate a chronic paranoid state. Misinterpretation of normal somatic features contributed to the chronic delusional paranoid psychosis.

D. Early Onset of Frightening Experiences

1. *Case WHJ*

This 38-year-old man had a chronic seizure disorder dating from infancy. Seizures would occur as frequently as three to four times a day.

a. Seizure Pattern and Psychosis. During the aura he would have an epigastric sensation, feeling afraid and strange, experience receptive aphasia ("I hear you but I don't hear anything you say"), and have auditory hallucinations. After a behavioral automatism, he would find himself in another location, sometimes at a considerable distance from his starting point. The postictal period would last 24–48 h and sometimes up to weeks. At this time, he would become increasingly suspicious and paranoid. These feelings became pervasive, developing into a chronic paranoid psychosis. With an increase in seizure frequency, he would have an exacerbation of the psychosis. When the latter was associated with self-neglect or aggressive behavior directed at his "enemies," so-called because he thought they had done something morally wrong, a mental hospital admission would be precipitated. Religion became an ongoing component of his thinking. He stated, "God's all through my body," and he pointed to his bilateral Dupuytren's contractures, which he believed to be stigmata. "The seizures help me in doing righteousness. I can live with them. I am one of the afflicted."

b. Comment. The frequency, vividness, and prolongation of the intrusion of unusual experiences into his life formed the basis of the development of the psychosis.

E. Seesaw Relationship Between Seizures and Psychosis

In only one patient did we see a seesaw situation which did not imply an alternation of seizures with psychosis, but, rather, was attributable to a third possibility—drug effect.

1. *Case WAC*

WAC, a 28-year-old woman, with onset of seizure disorder at 5 years, showed a seesaw relationship. Psychosis appeared for the first time at age 28 following unusual control of her previously frequent seizures by methsuximide. Illusions were prominent: "The bed is all wrong. There's no top sheet. Every time I try to get in, it's moved." When the drug was discontinued, the seizures recurred, and the psychotic symptoms subsided and did not recur during the subsequent 2 years.

a. Comment. Reduction of seizures did not seem to be the sole precipitant of psychosis. Following earlier implantation of a cerebellar stimulator (6 years previously), a 60-fold reduction (20–30 to 2–4/day) in the frequency of seizures had occurred without the appearance of psychotic ideation. In this case, it was postulated that a neuropharmacologic mechanism produced the apparent seesaw relationship: As anticonvulsant medication controlled seizures, it also caused behavioral symptoms of a psychotic kind.

VI. Response to Treatment Contributes to Our Understanding

1. Observation of paranoid episodes precipitated by a flurry of seizures.

Some of our patients are brought to the emergency room or the neurology or psychiatry unit because of a paranoid development. They are said to have clear or probable seizure activity at home and are brought to the hospital because of mental changes. Not infrequently, suppressed feelings come to the fore. For instance, GLE, longing into her 40s for a child, without fulfillment, would be convinced that she was pregnant, as well as having paranoid delusions relating to her questionably supportive husband. Such a patient usually will respond within a week with adjustment of antiepileptic drugs and, usually, administration of a neuroleptic, although whether that is really necessary is not clear. Once reality contact has been regained, the patient is ready again to accept her lot.

2. Effect of surgery on psychosis.

Our patients have not had recurrence of psychosis after temporal lobe surgery. We have seen only one psychosis *de novo* postoperatively. This occurred in KOM, a 32-year-old single man, who had a pre- and postoperative schizoid personality, with paranoid tendencies, but had never been clearly psychotic. After left anterior temporal resection, he was seizure-free for 4 years with no changes in his personality configuration. He was off all ASDs for 2 years. While visiting family out west, he became acutely paranoid and, after consultation by phone, was flown back to Medical College of Ohio (MCO). Scalp EEG revealed temporal lobe SE. The psychosis gradually receded as seizures were controlled and haloperidol administered.

VII. Summary

Patients with seizures arising in the temporal lobe area of the brain are especially prone to behavioral problems including psychosis. There is agreement that psychotic thinking may be associated with seizure activity or occur after termination of that activity: in the postictal period. Figures have ranged from 15 to 80%, to a prevalence not different from that in the general population. Complexity and confusion in regard to understanding the relationships between epilepsy and psychosis rest in the often inadequate or unclear description of the epileptic and psychotic phenomena in the individual patient. When all factors are considered and studied over time, there is a greater possibility that one can recognize the mechanisms and relationships pertaining to the psychosis. Elimination of seizures can be expected to be accompanied as well by elimination of major psychiatric difficulty such as psychosis, rather than the reverse.

It is clear that increased collaborative neuropsychiatric study of patients with epilepsy be made. This was emphasized in Trimble's currently published, "The Psychoses of Epilepsy" (Trimble, 1991) in which he says: "In contrast to the recognized need for proper psychometric evaluation prior to surgery, the necessity for adequate psychiatric assessment has not been realized in many surgical epilepsy centers. The number of good follow-up series of psychiatric data, considering the number of centers doing such procedures, is abysmal."

Problems in recognition of subtle ictal activity and the unpredictability with which psychosis follows similar biologic change (e.g., seizure events) may account for observations of a lengthy interval between onset of the seizure disorder and that of psychosis and of a seesaw mechanism between seizures and psychosis.

In addition, the notion of an unrelated alternating course between seizure activity and psychosis may be supplanted by the observation of an altered mood state and/or paranoid thinking directly preceding or following one or a series of diurnal or nocturnal seizures. Furthermore, the absence of major seizures is not equivalent to freedom from seizures, in view of the adverse behavioral effects of subtle seizure activity.

References

Betts, T. A. (1981). Depression, anxiety, epilepsy. *In* "Epilepsy and Psychiatry" (E. H. Reynolds and M. R. Trimble, Eds.), pp. 60–71. Churchill Livingstone, Edinburgh.

Bruens, J. H. (1974). Psychoses in epilepsy. *In* "Handbook of Clinical Neurology" (P. J. Vinken and G. W. Bruyn, Eds.), Vol. 15, pp. 593–610. North-Holland Publishing Co., Amsterdam.

Fenton, G. W. (1978). Epilepsy and psychosis. *Ir. Med. J.* **71,** 315.

Ferguson, S. M., Rayport, M., Gardner, R., Kass, W., Weiner, H., and Reiser, M. F. (1969). Similarities and mental content of psychotic states, spontaneous seizures, dreams, and responses to electrical brain stimulation in patients with temporal lobe epilepsy. *Psychosom. Med.* **31,** 479–498.

Flor-Henry, P. (1969). Psychosis and temporal lobe epilepsy. *Epilepsia* **10,** 363–395.

Gibbs, F. A. (1951). Ictal and non-ictal disorders in temporal lobe epilepsy. *J. Nerv. Ment. Dis.* **113,** 522–528.

Glaser, G. H. (1964). The problem of psychosis in psychomotor temporal lobe epileptics. *Epilepsia* **5,** 271–278.

Hill, D. (1953). Psychiatric Disorders of Epilepsy. *The Medical Press* **229,** 473–475.

Jackson, J. H. (1875). 'On temporary mental disorders after epileptic paroxysms,' in West Riding Lunatic Asylum Medical Reports, Vol. 5, pp. 109–129. *In* "Selected Writings of John Hughlings Jackson, Vol. I, On Epilepsy and Epileptiform Convulsions" (J. Taylor, Ed.), 1958, 119–134. Basic Books, New York.

Kraeplin, E. (1923). "Psychiatrie." Johann Ambrosius Barth, Leipzig.

Landolt, H. (1958). Serial electroencephalographic investigations during psychotic episodes in epileptic patients and during schizophrenic attacks. *In* "Lectures on Epilepsy" (A. M. Lorentz de Haas, Ed.), pp. 91–133. Elsevier, Amsterdam.

Parnas, J., and Korsgaard, S. (1982). Epilepsy and psychosis. *Acta Psychiatr. Scand.* **66,** 89.

Perez, M. M., and Trimble, M. R. (1980). Epileptic psychosis: Diagnostic comparisons with process schizophrenia. *Br. J. Psychiatry* **137,** 245–249.

Pond, D. A. (1957). Psychiatric aspects of epilepsy. *J. Indian Med. Profession* **3,** 1441–1451.

Post, R. M., *et al.* (1982). Carbamazepine, temporal lobe epilepsy, and manic-depressive illness. *Adv. Biol. Psychiatry* **8,** 117–156.

Reynolds, E. H. (1968). Epilepsy and schizophrenia: Relationship and biochemistry. *Lancet* **1,** 398–401.

Rice, J. L. (1985). *In* "Dostoevsky and the Healing Art: An Essay in Literary and Medical History," p. 84. Ardis, Ann Arbor.

Slater, E., Beard, H. W., and Glithero, E. (1963). The Schizophrenia-like psychoses of epilepsy. *Br. J. Psychiatry* **109,** 95–150.

Stevens, J. (1966). Psychiatric implications of psychomotor epilepsy. *Arch. Gen. Psychiatry* **14,** 461–471.

Symonds, C. (1962). Discussion. *Proc. R. Soc. Med.* **55,** 4–5.

Temkin, O. (1971). *In* "The Falling Sickness," pp. 4–5, 15, 151–152, 361. Johns Hopkins Press, Baltimore.

Toone, B. K. (1981). The psychoses of epilepsy. *In* "Epilepsy and Psychiatry" (E. H. Reynolds and M. R. Trimble, Eds.), pp. 113–137. Churchill Livingstone, Edinburgh.

Trimble, M. R. (1977). The relationship between epilepsy and schizophrenia: A biochemical hypothesis. *Biol. Psychiatry* **12,** 299–304.

Trimble, M. R. (1991). "The Psychoses of Epilepsy." Raven Press, New York.

Trimble, M. R., and Meldrum, B. S. (1979). Monoamines, epilepsy and schizophrenia. *In* "Biological Psychiatry Today" (J. Obiols, E. Ballus, M. Gonzales, and J. Pujol, Eds.), pp. 470–475. Elsevier, Amsterdam.

Trimble, M. R., and Perez, M. M. (1982). The phenomenology of chronic psychoses of epilepsy. *Recent Adv. Biol. Psychiatry* **8,** 98.

SECTION IV

OUTCOME OF TEMPORAL LOBECTOMY

MEMORY FUNCTION AFTER TEMPORAL LOBECTOMY FOR SEIZURE CONTROL: A COMPARATIVE NEUROPSYCHIATRIC AND NEUROPSYCHOLOGICAL STUDY

Shirley M. Ferguson,* A. John McSweeny,*,† and Mark Rayport‡,§

*Department of Psychiatry, University of Toledo College of Medicine (formerly Medical College of Ohio), Toledo, Ohio 43614, USA
†Department of Neurology, University of Toledo College of Medicine (formerly Medical College of Ohio), Toledo, Ohio 43614, USA
‡Department of Neurological Surgery, Medical College of Ohio, Toledo, Ohio 43614, USA
§Department of Radiology, Medical College of Ohio, Toledo, Ohio 43614, USA

This chapter deals with often differing data and prognosis provided by the neuropsychiatrist and the neuropsychologist. It also reflects the general psychology literature which raises the question as to whether laboratory testing of memory function truly reflects memory function in everyday life.

Current practice in surgical epilepsy centers is a preoperative neuropsychological evaluation, particularly of memory function. This demand was stimulated by the occurrence of a devastating amnesia which occurred in the 1950s in the wake of temporal lobe surgery. The amnesia was attributed to hippocampal dysfunction, either from removal of both hippocampi at surgery or from removal of one hippocampus when the other was dysfunctional. Subsequently, to avoid a memory calamity, testing of the hippocampus to remain has become mandatory prior to excision.

The literature is replete with prediction reporting of memory impairment post temporal lobectomy, even though the remaining hippocampus has been

INTERNATIONAL REVIEW OF NEUROBIOLOGY, VOL. 76
DOI: 10.1016/S0074-7742(06)76005-6

0074-7742/06 $35.00

reported as functional, by preoperative memory testing. The general psychology literature also raises the question as to whether laboratory testing of memory function truly reflects memory function in everyday life.

Our longitudinal study of epilepsy patients who have undergone temporal lobectomy at the Medical College of Ohio has emphasized utilization of neuropsychiatric evaluation as well as neuropsychological testing. Data and prognosis reported by the neuropsychologist are frequently in contrast with neuropsychiatric assessment, which provides subjective reports by the patient as well as review of performance-demanding memory ability, in the spheres of education and work. Comparisons of neuropsychological and neuropsychiatric data were expressed in terms of concordant or discordant findings.

Data for this study were gathered between 1996 and 1998.

I. Introduction

The consequences for memory function of unilateral temporal lobectomy for control of intractable temporolimbic seizures (TLmS) have been investigated extensively with neuropsychological test batteries in a vast output of publications that include groundbreaking contributions.

In the 32-year experience of the epilepsy comprehensive program (ECP) of the Medical College of Ohio, we noted at its multidisciplinary conferences a recurring dissimilarity between the patient's self-assessment of memory function after temporal lobectomy for seizure control (TLX) as reported to and evaluated by the neuropsychiatrist and memory tests performed in the neuropsychological laboratory. The neuropsychiatrist more often reported an improvement in memory function as reflected in the patient's descriptions of subjective experiences and day-to-day transactions whereas the neuropsychologist more often reported decline in memory scores. These observations led us to compare the pre- and postoperative results of neuropsychiatric assessment of memory ability with those of structured neuropsychological measures of memory function in the neuropsychology laboratory.

II. Literature

Persistent inability to lay down new memories and extensive retrograde amnesia were encountered unexpectedly by Penfield in the late 1940s after unilateral removal of hippocampus and adjacent medial temporal cortices in a patient on whom he had previously performed anterior temporal corticectomy and amygdalectomy without achieving seizure control. Penfield postulated that the nonoperated hippocampus was nonfunctional. The result of the unilateral temporal lobectomy was therefore equivalent to bilateral excision, leaving the patient without hippocampal function (Milner and Penfield, 1955; Penfield and Milner, 1958). Devastating amnesia ensued from one-stage bilateral removal of the medial temporal lobe structures in Scoville's Case HM (Corkin, 1992; Scoville and Milner, 1957). Observations on Case HM were seminal of extensive and brilliant investigations of the role of the hippocampus in memory function (Milner, 1954).

The alarming hippocampal amnesia syndrome that curtailed normal life created a situation in which extreme caution was subsequently exercised by the epilepsy neurosurgeon who wished to safely continue the valuable surgical treatment for intractable temporal lobe epilepsy. The neuropsychologist developed test batteries which improved the patient's safe passage through temporal lobectomy for seizure control. The neuropsychologist became an established member of the epilepsy surgery team (Rayport and Ferguson, 1996). Refinement in presurgical preparation was obtained when the Wada test that had addressed preoperative cerebral lateralization of language was extended to memory function (Milner *et al.*, 1962; Wada and Rasmussen, 1960).

Memory deficits demonstrable before and after TLX were material specific, left temporal lobe involvement being associated with verbal and right temporal with nonverbal remembering (Meyer and Yates, 1955; Milner, 1954, 1975). Blakemore and Falconer (1967) stated that, although there might be some memory impairment after TLX, in 3 years time memory usually improved. Improvement over time was also observed by others (McGlone and Wands, 1991; Novelly *et al.*, 1984). Milner (1975) stated that even though there were abnormalities on psychological tests of memory function after TLX, "they would not interfere significantly in the patient's everyday life."

In the 1970s, research psychologists were remarking on the complexity of memory function and questioned whether tests in the laboratory truly reflected the person's memory ability in everyday life. Wilkins and Baddely (1978) attempted to simulate real-life situations which required adequate memory function. Thus, referring to the epileptic patient dependent on taking medication regularly, they supplied the patient with a box that contained a clock system which recorded the times at which a button on the outside was operated. The patient was instructed to press the button four times a day, to reproduce the pattern of a medication schedule. The authors found that although the patient might have been good at this simulated drug taking, he might have a poor memory for a list of words. The authors concluded that, "There is no indication of the behavior that leads to remembering or forgetting." Wilkins (1986) commented that, "In the laboratory, one can measure remembering to do things without a cue, but there are no studies as to the predictive validity for everyday life." He further stated that, "Nobody knows how people remember without being reminded even though remembering to do something without being reminded is one of the most important aspects of memory function in everyday life. It remains to be seen which types of memory tests will be most sensitive to the particular memory difficulty the patient experiences."

Attempts to resolve measurement problems were made by developing tests which might more clearly reflect patients' memory ability in everyday life. English psychologists developed and later revised the Rivermead Behavioral Memory Test which emphasized inquiry into everyday memory ability. The test was declared "promising" (Wall *et al.*, 1994). Wilkins (1986) stated that, "the way memory is examined in the clinical context needs to be adapted so as to take into account limitations inherent in the psychometric tradition from which the work stems." In addition to standardized psychological tests, computerized testing was utilized by Ivnik *et al.* (1993). These authors concluded that, although this approach seemed to provide more meaningful results, "the impact of memory problems in daily life must be assessed," inferring that laboratory testing of memory must be correlated with or complemented by direct observation of everyday functioning.

In the 1980s and 1990s, attention was called to the very question of whether memory impairment manifested in laboratory testing was of concern for the patient after TLX. Milner (1975) had stated that, "these deficits tend to be mild and to interfere little if at all with the patient's daily life." Bennett-Levy *et al.* (1980) studied patients who had undergone TLX. The authors (Novelly *et al.*, 1984; Phillips and McGlone, 1995) compared the deficits manifested in laboratory tests and the self-report of real-life memory skills. They concluded that, "In the absence of real-life measures, little is known about the consequences of organic damage for day-to-day living." Several authors found that memory function improved if seizure control was attained. They emphasized the need to study the individual patient and found gains as well as losses in cognitive functions. The latter authors provided a practical conclusion that "restructuring the surgical procedure for temporal lobe excision to avoid memory decline does not seem warranted."

In the 1990s, more attention was paid to anatomical/pathological findings and to the use of imaging techniques in patients undergoing TLX. Seemingly contradictory findings resulted. Hermann *et al.* (1994) tested 20 patients before and 6 months after TLX. They compared subjective and objective reports of memory function with the degree of hippocampal sclerosis. They found greater decline in episodic memory in patients with normal or mildly sclerotic hippocampi. Other authors (Helmstaedter and Elger, 1996) differentiated the effect of surgery on mesial or neocortical temporal structures. They found that left anterior TLX affected verbal acquisition and recognition more than consolidation/retrieval.

Comparison of subjective and objective memory changes after TLX showed differences in incidence and in significance. Sawrie *et al.* (1999) found the two unrelated. They stated that although objective decline may be common, there is little noticeable impact on everyday life. Subjective complains were infrequent. If they did occur, the authors concluded that they served as a marker for emotional distress.

Another logical question was how demand on memory affects the manifestations of memory decline. Contrary to anticipation, that the higher the demand, the more the subjective complaints were to be expected, the opposite was noted. Gleissner *et al.* (1998) found that the lower the demand, the greater the subjective complaint of decreased memory ability.

Later, simulated tasks were devised for administration in the laboratory. For instance, subjects were asked to watch a film while following instructions to monitor the baking of a cake. Epileptic patients who had had large excisions of the left hippocampus did poorly on such tests. Small removals did not produce a deficit.

The various discrepancies in correlating memory function with brain areas can be understood in the light of the recognition, beginning in the 1980s, that there are multiple forms of memory and that several brain areas, besides the hippocampus, are involved with memory function (Squire, 1994). Thus, one would have to question the validity or meaningfulness of associating brain changes on imaging studies with cognitive function. The complexity of brain circuits and of memory function and their relationships most likely accounts for discrepancies and contradictory findings and for claims in the literature about memory and its relationship to brain areas. The impact of emotional pressures and of context has also been shown to have a role in memory ability.

III. Clinical Material

A. Study Population

The study population consisted of 22 patients who were seizure-free since TLX, ranging in age from 14 to 54 years at the time of operation. Thirteen were females. Neurosurgical and neurological information has been published earlier (Rayport and Ferguson, 1996). The patients were studied longitudinally prior to TLX from 1 month to 9.8 years, average 6.1 years. Postoperative follow-up ranged from 1.4 to 21 years (mean duration 6.1 years, median 11.4 years). Final assessment of each patient was made in the last years of postoperative follow-up. Patients were recalled for this study between 1996 and 1998.

B. Methodology

1. *Neuropsychiatric Methodology*

Preoperative neuropsychiatric evaluation was initiated when the patient with intractable seizures was referred to the multidisciplinary epilepsy team because of medication-resistant seizures, or to the neuropsychiatrist because of emotional/behavioral problems as well as seizures. Preoperative neuropsychiatric evaluation was continued until TLX.

The preoperative neuropsychiatric assessment consisted of an initial structured interview and examination followed by longitudinal observations at each visit to the multidisciplinary epilepsy clinic and during seizure-related hospitalizations. The initial interview included a detailed seizure history. Questions directed specifically at memory function were structured and open ended. Additionally, memory function was assessed by its reflection in the patient's ability in dealing with everyday transactions. These data were reviewed and possibly extended during interviews with a family member (Rayport and Ferguson, 1996). As each patient made subsequent scheduled visits to the multidisciplinary epilepsy clinic, s/he was seen by the neurosurgical or neurological epileptologist and the neuropsychiatric epileptologist, the clinical nurse specialist, and a social worker/counselor if indicated. Further neuropsychiatric evaluation during these contacts included observations of memory function.

After TLX, patients were followed by the same multidisciplinary team. Longitudinal postoperative observation ranged from 1.4 to 21 years (mean duration 6.1 years, median 11.4 years) when the final formal structured postoperative assessment of memory function was undertaken for this study, which was approved and monitored by the Research and Grants Administration at Medical College of Ohio.

The protocol was as follows: Personal phone contact was made by the neuropsychiatrist or clinical nurse specialist with each patient to propose participation in the study. Patients were told that we were contacting other patients who had had similar surgery, that we were interested in the overall outcome of the surgery, and, particularly, in their memory ability.

When the patient arrived in the waiting room, s/he was asked to complete an open-ended, broadly directed form about overall outcome. Questions covered current status in regard to seizures and medication, living arrangements, interpersonal relationships, accomplishments and/or problems at school or work, and any emotional problems. An identical questionnaire was completed by a family member.

At the session with the patient, which lasted about 1 h, there were two aspects of the inquiry.

1. Patient's spontaneous report in regard to overall life and, specifically, current memory function. Patient was asked if s/he had noted any contrast between ability before and after surgery.

2. Structured inquiry about different types of memory activity: function in school or at work; keeping track of possessions; recalling occurrence and/or content of telephone messages for someone else; sharing ongoing experiences; retaining reading matter, content of TV programs, movies, and so on; and items of personal and general past. In each category, patient was asked to compare his ability before and after surgery.

Data obtained by preoperative history from patient and from a family member were compared with those resulting from the postoperative assessment to determine whether memory function was the same, worse, or better after seizure-relieving surgery. In each case, the general life situation before and after TLX was considered an excellent measure of the effect of seizure eradication on memory function in everyday life.

2. *Neuropsychologic Methodology*

The patients all received a comprehensive neuropsychological test battery as part of their total evaluation within the ECP. The test battery was designed to measure general intellectual functioning, spatial abilities, language, motor skills, and attention in addition to memory. The battery varied in terms of the specific instruments employed given that the study took place over a relatively long time span. Several instruments were added as they became available for use and others were deleted as they became obsolete. In particular, this applies to the Wechsler Intelligence and Memory Scales. Early in the study, the Wechsler Adult Intelligence Scale (WAIS) was employed. As the revised (WAIS-R) and third versions became available, these were substituted for the first version. Similarly, the original version of the Wechsler Memory Scale (WMS), supplemented by delayed recall trials of story memory and figural memory, was used first and then replaced by the revised and third versions. Patients received the same version of the test at postsurgical visits that was administered during the presurgical visit.

In addition to calculating the usual scores and indices from one of the versions of the WMS, the neuropsychologist also made blind clinical ratings of memory functioning for general memory, verbal memory, and nonverbal memory. These ratings used a six-point scale where 0 = above normal, 1 = normal, 2 = borderline, 3 = mild impairment, 4 = moderate impairment, and 5 = severe impairment. In addition, the neuropsychologist made a rating of change from presurgical to postsurgical assessment as improved impaired or unchanged. Finally, the neuropsychologist made a descriptive judgment of patient's prognosis for future functioning based on the memory testing.

3. *Methodology of Comparison*

The global neuropsychiatric findings reflecting actual memory performance in everyday life were compared with the prognostic conclusions about memory function based on neuropsychological testing.

IV. Results

A. Subjective Evaluation of Memory (Likert Scale and Verbal Comments)

After left TLX, patients reported that their memory was not changed (no problems before or after surgery) or improved ("very much better" or "better"). In the following table (Table I), pre- and postoperative comments about memory function are presented.

After right TLX, patients also reported that their memory was improved or not changed. Three stated that their memory function was "very much better," and two that it was "better." Six patients reported no problems before or after surgery. However, two patients found that after surgery, their memory was better than it had been ("It is easier to remember," or "My memory is beautiful, better than it ever was"). See Table II.

TABLE I
Postoperative Subjective Evaluation After Left TLX (Likert Scale)

Demographics			Likert score	
Patient	Age at TLX	Date TLX	Memory	Function
ARM	41	1987	Much better	Much better
COP	36	1987	Better	Much better
KIT	54	1992	Much better	Much better
BASA	35	1980	Much better	Much better
CRT	17	1988	Better	Much better
FRS	14	1980	Better	Much better
KIJ	30	1983	Better	Much better
LAK	25	1993	Much better	Much better
SMA	32	1985	Much better	Much better
SWC	41	1982	Better	Much better
WEN	30	1982	Much better	Much better

TABLE II
POSTOPERATIVE SUBJECTIVE EVALUATION AFTER RIGHT TLX (LIKERT SCALE)

Demographics			Likert score	
Patient	Age at TLX	Date TLX	Memory	Function
BAS	21	1977	Better Still some problems "Have to look things up."	Much better
COR	36	1993	No change "No problem before or after."	Much better
MYD	22	1982	Better No problem before Postoperation: "Easier to remember"	Much better
BRB	26	1985	Much better	Much better
BRJ	24	1991	Much better "Can do things without being retold"	Better
CAK	27	1974	No change (no problem before or after)	Much better
LAS	26	1993	Better No complaint before. Postoperation: "Memory beautiful—better than it ever was."	Much better
NIH	39	1986	Much better Preoperation: "Memory was killing me." Postoperation: "Reading easier and more quickly and increased comprehension."	Much better
PAB	27	1975	No change (no problem before or after)	Much better
RUD		1979	No change	Much better
SHB	24	1981	No change Satisfactory before and after Remembers shared experiences; content of reading and TV programs	Much better

B. EDUCATIONAL EXPERIENCE AND ACCOMPLISHMENT

1. *Prior to Left TLX*

One patient achieved a college degree prior to her chronic TLmS. Five patients earned a high school diploma. Among them, four had particular difficulty with verbal tasks such as understanding and retaining what they read.

2. *Prior to Right TLX*

One patient had a college degree. Six graduated from high school. No information about grades for two. The other four had either B to low C averages, or "problems," or were in learning disabled classes. One had earned a general educational development (GED). Three left school, either on their own or at the request of school authorities.

3. *Post TLX*

A trend to improvement in ability/performance is evident as shown in Table III.

C. Work History

1. *Prior to Left TLX*

Three patients had been employed: One was a struggling elementary school teacher, one was a packer in a factory, and one had sheltered work (work fare) in the welfare department. Six patients had been unemployed from 2 to 10 years. (Two were in high school, not doing well.)

2. *Prior to Right TLX*

Four patients had been employed: Only one was satisfied with his job as a packer in a grocery store. One considered his stockroom work "boring"; one was dissatisfied with the constant supervision provided by family at home and in the family business; one did menial jobs. One kept busy pursuing college courses (in a master's degree program), although memory impairment (worsened by seizures) caused him to struggle, not do well, and frequently to drop a course. Two were home bound: One since being expelled from school at age 13 and one was self-isolated at home. One had been in a psychiatric hospital for 7 years.

3. *Post TLX*

All patients improved: They advanced or improved in their previous employment, became gainfully employed, assumed new responsibility at home, became a volunteer in the community, or achieved independent living after long hospitalization.

Comparison of details of the pre- and postoperative work status in the two TLX groups is also shown in Table III.

V. Comparison of Neuropsychiatric and Neuropsychological Evaluation of Memory Function After TLX

Neuropsychiatric assessment of memory function was based on cognitive testing, subjective reports, and objective evidence of performance in life tasks of

TABLE III

Functioning in Everyday Life After TLX: Education and Work

Demographics			Education		Work	
Patient	Age at TLX	Date TLX	Preoperation	Postoperation	Preoperation	Postoperation
Left TLX						
ARM	41	1987	HS dropout		Hospital ward clerk. Unemployed 3 years	Filing clerk in store 4 years. Cared for husband 2 years. Office work past 4 years
SMA	32	1985	HS dropout		Homebound, constant supervision	Baby sitter. Manages household. One child (1992)
COP	36	1987	HS diploma—did not do well	Obtained a real estate license	Homebound	Employed as bookkeeper
KIJ	30	1983	HS diploma	Not applicable	Packer in factory	Promoted to inspector
SWC	41	1982	HS diploma	Did well in course in maintenance	Unemployed several years	Maintenance worker in hotel
WEN	30	1982	HS diploma	College courses	Sheltered work (welfare office)	Same, with promotions
KIT	54	1992	HS diploma	Training in business management	Unemployed	Ran own business: concession in police department
LAK	25	1993	HS diploma (special classes)	Not applicable	Dismissed from work (fast food) because of seizures. On Supplemental Security Income	Food service in nursing home, 6 years
CRT	17	1988	In HS, not doing well	Grades improved. HS diploma	Not applicable	Assistant manager in photo department. Married, two children
FRS	14	1980	In HS, not doing well	Grades improved. HS diploma. Trained in preschool child care	Not applicable	Employed 8 years (preschool teacher). Married, one child

(*Continued*)

TABLE III (*Continued*)

Demographics			Education		Work	
Patient	Age at TLX	Date TLX	Preoperation	Postoperation	Preoperation	Postoperation
Left TLX (*cont.*)						
BASA	35	1980	BA (before onset of chronic seizure disorder)	Not applicable	Elementary school teacher. Problems in remembering names, ongoing experiences, and so on	Same but improved performance. Knows names of children at end of first day. Retains better in general. More interests
Right TLX						
BAS	21	1977	HS graduate, 1974 (B–C average). Failed college courses	Community college, X-ray technology. 1982–1985	Stockroom "boring"	X-ray technician
BRB	26	1985	HS graduate, 1977 (cognitive and social difficulties)	Medical assistant course, 1989 (9 months)	Sheltered work in family business	Independent: housewife, mother, medical assistant course, sings in chorus
BRJ	24	1991	HS diploma (LD classes 5–10). Memory impaired	Not applicable	Packer in grocery store	Same location with additional assignments plus farmhand for 8 farmers
COR	36	1993	Left school: difficulty with teachers. GED	Pre-BA nursing course at university	Unemployed	Factory work (while attending school part-time)

LAS	26	1993	HS graduate, 1984 (low C)	Associate degree in computer assisted manufacturing (1996) dean's list	Unemployed 4 years	Computerized tool and dye worker
MYD	22	1982	Left school, 1977	Courses at community college	Employed in heating and air conditioning company	Advances. Training others. Considering opening his own business
NIH	39	1986	BS, Texas, 1974. MA program Ohio	Courses "to keep occupied," successful	Unemployed. Never steady job	Laboratory assistant 1989–1991. Research laboratory technician 1992–1995. Substitute teacher 2000
SHB	24	1981	At 13 (7th grade): asked to leave school	Failed GED	Homebound	Managed household (mother ill—1989). Food service in nursing home 1 year, currently working in deli
CAK	27	1974	HS diploma	Nurses aid course	Homebound. Two children	Volunteer: Welcome Wagon. Nurses aid (nursing home). Works for husband
PAB	27	1975	Grammar school	Classes at YWCA	Psychiatric hospitalization for 7 years	Writes poetry, articles. Volunteers at epilepsy center. Crochets (sells). Active at church. Remarried—cares for husband on dialysis
RUD		1979	HS diploma	Nursing school (LPN), graduated 1983	Unskilled work: cuts grass, and so on	LPN

education and working. Neuropsychological testing of memory function led to prognostic statements about the probable outcome of memory function after TLX. Neuropsychiatric and neuropsychologic findings were concordant in 7 patients and discordant in 15. Tables IV and V provide evidence for this conclusion.

Subjective and objective neuropsychiatric findings reflected everyday memory ability and were compared with neuropsychological prognosis.

A. Left TLX

In three cases, neuropsychiatric and neuropsychologic findings were concordant; in eight patients, discordant.

1. *Examples of Concordance*

One female patient, previously unemployed, found office work in a department store where she remained for 4 years until her husband's terminal illness caused her to leave to care for him. After his death 2 years later, she returned to work and has been employed for over 4 years. The neuropsychological prognosis: "No restrictions on quality of life."

A male patient, preoperatively unemployed for 4 years, started an independent food service concession in a government building. Neuropsychological prognosis: "Likely to manage memory tasks adequately."

2. *Examples of Discordance*

On the other hand, among the eight patients whose assessments were discordant, there were such findings as:

An elementary school teacher, who had been struggling to perform prior to surgery, showed obvious change for the better. She was promoted to teach upper grades, could complete tasks more quickly and efficiently, and reported proudly that she could now learn pupils' names on the first day of school (previously prolonged and effortful).

The neuropsychologist reported that, "she has prospects for jobs not entailing retention of verbal instructions. She will require retraining in note taking etc."

A former high school-failing female student greatly improved her performance when she returned to school after operation. After graduation from high school, she completed a course in early child care and has been employed for 8 plus years as a preschool teacher.

The neuropsychologist prognosticated that, "She may have trouble recalling verbal information on a day-to-day basis."

B. Right TLX

In three cases, neuropsychiatric and neuropsychologic prognosis was concordant, in eight cases, discordant.

1. *Examples of Concordance*

One male patient who had been performing "boring" stockroom work finally decided to become an X-ray technician. He passed his course and has been employed in the field for 15 years. He explained his relative slowness in that "I may have to look things up." The neuropsychologist concurred: "He will have slight difficulty with memory. Minimal compensatory strategies will be required."

Preoperative, a male patient had been laid off because of seizures. On returning, he did well, transferred to another location to express his independence (first job obtained with help of father, a long-time employee in that location). He took courses, was promoted, began to train others, and is thinking of starting his own business. The neuropsychologist agreed: "No impairment expected."

2. *Examples of Discordance*

One male patient had worked as a packer in a grocery store. On returning, he was given other assignments—"wherever needed." He also worked as a farm-hand for eight farmers in a two-county area.

The neuropsychological conclusion: "He will have a mild impairment of non-verbal memory. He will have minimal interference with day to day life."

Another male patient, a tool and dye maker, had not worked for 4 years preoperatively. After operation, he attended a community college to take a computerized tool and dye design course. He was on the dean's list. He returned to the company where he has been a valued employee for several years, especially since he is the only employee so trained.

The neuropsychological prognosis: "He is likely to have considerable problem with non-verbal memory."

Table IV and Table V present the concordance/discordance data.

VI. Discussion

Memory research has been actively pursued in the neuropsychological literature since the 1950s, especially in relation to the effect of temporal lobectomy on memory function. Because of the threat of devastating memory loss as described in the historic case of HM by Milner (1954), testing of memory function preoperatively became mandatory in surgical epilepsy centers. As the central behavioral theme in the study of patients with surgical epilepsy, memory testing called forth neuropsychology as the chief behavioral discipline involved in studies of patients before and after temporal lobe surgery for medically intractable seizures.

In the 1970s, after focusing for two decades on cognitive function after surgery for seizure control, neuropsychological investigators began to point to the need to address the larger sphere of quality of life. Questionnaires were developed which, hopefully, would adequately reflect the patient's course. Inquiry utilized new techniques as they became available. Thus, computerized programs were

TABLE IV

COMPARISON OF NEUROPSYHIATRIC AND NEUROPSYCHOLOGIC DATA LEFT TLX: CONCORDANCE/DISCORDANCE

			Neuropsychiatry		Neuropsychology
			Comparison of pre- and postoperative function		
Patient	Date	Age	Preoperation	Postoperation	Prognosis postoperation
Concordant					
ARM	1987	41	Unemployed 3 years	Office work in store 4 years. Cared for ill husband 2 years. Currently, office work 4 years	No restrictions on quality of life
COP	1987	36	Homebound	Bookkeeping jobs. Real estate license. May request repeat of instructions	Will need to use some memory aids but major restrictions in everyday life not necessary
KIT	1992	54	Unemployed	Owns own business (food service in government building) and volunteer in hospital	Likely to manage memory tasks adequately
Discordant					
BASA	1980	35	Elementary school teacher with problems due to impaired memory and interference by seizures	Improved performance. Knows pupils' names after first day	Prospects for jobs not entailing retention of verbal instructions. Will require retraining in note taking, and so on

CRT	1988	17	In high school. Poor grades.	Grades improved. Employed as assistant manager in photo department. Helped in husband's business after birth of child	May require memory assists to compensate for decreased recall. Memory difficulty will not interfere with daily living
FRS	1980	14	In high school. Poor grades.	Grades improved. Preschool day care training and work (8 years)	May have trouble recalling verbal information on day-to-day basis
KIJ	1983	30	Packer in factory	Inspector. Set up machines	Significant impairment in memory in ALD. Verbal especially impaired
LAK	1993	25	Dismissed from fast food jobs because of seizures. Unemployed—on SSI	Nursing home in (food service) 6 years	Will affect life to a mild degree. May need to use memory aids
SMA	1985	32	Unemployed 10 years	Babysitting (several children). One child 1992	Memory assist needed to mild significant degree
SWC	1982	41	Unemployed because of seizures	Maintenance work at hotel 6 years	Will need assistance with memory-related tasks.
WEN	1982	30	Sheltered work (welfare department)	Same location—promotions	Will need to rely on some memory aids but difficulty in every day life mild

TABLE V

COMPARISON OF NEUROPSYHIATRIC AND NEUROPSYCHOLOGIC DATA RIGHT TLX: CONCORDANCE/DISCORDANCE

			Neuropsychiatry		Neuropsychology
			Comparison of pre- and postoperative function		
Patient	Date	Age	Preoperation	Postoperation	Prognosis postoperation
Concordant					
BAS	1977	21	Boring work (stockroom). College courses—failed	Training as X-ray technician. Employed as certified X-ray technician in hospital/office 13 years. Somewhat slow ("may have to look things up")	Slight difficulty with memory. Minimal compensatory strategies required
COR	1993	36	Unemployed 6 years	Employed and attending college (prenursing)	Likely to function normally for the most part
MYD	1982	22	Laid off because of seizures	Excellent job—advancing, taking courses, training others, considering starting own business	No impairment expected
Discordant					
BRB	1985	26	Sheltered in family business	Left job. Independent. Medical assistant course. Active in singing group. One child	Minimal problem with tasks requiring visual memory (locations on map)
BRJ	1991	24	Grocery store—packer	Same and also farmhand for 8 farmers in two-county area	Mild impairment of nonverbal memory. Minimal interference with day-to-day life

CAK	1974	27	Isolated. Feels shame at having seizures	Active in community, for example, Welcome Wagon; knits (complex patterns)	May have occasional trouble with spatial tasks
LAS	1993	26	High school. Low C average. Apprentice (tool and die). Unemployed several years	Community technical college: dean's list, associate degree. Excellent job in computer design (tool and dye)	Likely to have considerable problem with nonverbal memory
NIH	1986	39	Eternal student (taking and dropping courses)	First steady jobs in life. Laboratory technician 2 years and research laboratory 7 years	Likely to have significant difficulty in learning new information
PAB	1975	27	Psychiatric hospital 7 years	Independent living. Managing ill husband, schedule, diet, appointments, and so on	Mild impairment of memory in day-to-day activities
RUD	1979	24	Odd jobs	School licensed LPN. Employed 15 years in nursing home	Limitations in quality of life due to difficulty with memory. Will need memory aids
SHB	1981	24	Reversed day/night. Isolated	Took over household (ill mother). Employed: food service in nursing home 1 year; deli, currently	Impairment of day-to-day function. Will need memory aids

developed which, it was thought, would more clearly delineate the patient's success or failure in life. However, in several papers, the authors pointed to the need for better real-life assessment.

It is of interest that at the same time, in the general psychological literature, a similar question was being asked: How well do test findings of memory function obtained in the laboratory reflect the individual's memory ability in everyday life? Various approaches to testing were developed to more closely imitate life. In the end, there was agreement that memory is very complex and that an individual might perform well in one test and not in another and that context, the nature of the task, and other factors might affect the person's memory ability. Furthermore, since the 1980s, it has been recognized that there are multiple forms of memory and that several brain areas, besides the hippocampus, are involved with memory function. A true, meaningful quantification is hard to define.

Altogether the literature, both of general psychology and of neuropsychology, emphasized the complexity of memory function and its relation to context, to task involved, to anatomical location, and so on. Great effort was made to simulate real life.

In our work, dating from the 1960s, we decided to do this very thing. Although neuropsychological testing was continued in our ECP, neuropsychiatry became the most prominent behavioral discipline. With a phenomenological orientation, we conducted a prospective, longitudinal study to find out what the patient could do before and after surgery and how he evaluated himself. We observed life, rather than trying to simulate it. We assumed that memory ability would be key to adequate performance of everyday tasks, particularly in the prominent areas of life, education, and work. In order to understand how the patient experienced his memory ability, we utilized the Likert scale to gain comparative data.

With the subjective report about memory and a history of the patient's activities of everyday life over time, we then compared the latter to the postoperative neuropsychological prognosis, as to what laboratory tests predicted as to how the patient would function after surgery. We found that our conclusions were more often discordant, that is, prediction of difficulties by the neuropsychologist might contrast sharply with the patient's actual performance and accomplishments.

Reasonable questions to ask were whether and how seizures and cognitive dysfunction impede the patients before surgery. Once relieved of their seizures, were patients able to make their lives more meaningful and satisfying? We found that not only did they give positive reports about their memory ability, but the majority provided proof of its adequacy by studying and/or working at a higher lever than possible before surgery. Society often recognized their improvement by giving them more complex, more responsible assignments.

Careful longitudinal interdisciplinary studies characterized our work. Thus, we came to know our patients and their handling of their everyday life situations. We felt that the result should not be a test result but a description of the patient's life. Our role was to help the patient utilize and maximize his cognitive abilities now that he was no longer in the grip of unpredictable and disruptive seizures.

VII. Conclusions

Considering the difficulties of achieving a meaningful measure of memory function in the laboratory and through simulated tasks in and outside the laboratory, it seemed logical to use life performance as the most useful indicator of the competency of everyday memory function, which is, after all, the long-standing goal of memory research in patients who undergo TLX.

References

Bennett-Levy, A. J., Polkey, C. E., and Powell, G. E. (1980). Self-report of memory skills after temporal lobectomy. The effect of clinical variables. *Cortex* **16,** 543–547.

Blakemore, C. B., and Falconer, M. A. (1967). Long-term effects of anterior temporal lobectomy on certain cognitive functions. *J. Neurol. Neurosurg. Psychiatry* **30,** 364–367.

Corkin, S. (1992). Lasting consequences of bilateral medial temporal lobectomy: Clinical course and experimental findings in H.M. *In* "Frontiers in Cognitive Neuroscience" (S. M. Kosslyn and R. A. Andersen, Eds.), p. 516. The MIT Press, Cambridge.

Gleissner, U., Helmstaedter, C., Friske, A., and Elger, C. E. (1998). The performance-complaint relationship in patients with epilepsy: A matter of daily demands? *Epilepsy Res.* **32,** 401–409.

Helmstaedter, C., and Elger, C. E. (1996). Cognitive consequence of two-thirds anterior temporal lobectomy on verbal memory in 144 patients: A three month follow-up study. *Epilepsia* **37**(2), 171–180.

Hermann, B. P., Wyler, A. R., Somes, G., Dohan, F. C., Berry, A. D. III, and Clement, L. (1994). Declarative memory following anterior lobectomy in humans. *Behav. Neurosci.* **108,** 3–10.

Ivnik, R. J., Malec, J. F., Sharbrough, F. W., Cascino, G. D., Hirschorn, K. A., Crook, T. H., and Larrabee, G. I. (1993). Traditional and computerized assessment procedures applied to the evaluation of memory change after temporal lobectomy. *Arch. Clin. Neuropsychol.* **8,** 69–81.

McGlone, J., and Wands, K. (1991). Self-report of memory function in patients with temporal lobe epilepsy and temporal lobectomy. *Cortex* **27,** 19–28.

Meyer, V., and Yates, A. J. (1955). Intellectual changes following temporal lobectomy for psychomotor epilepsy. *J. Neurol. Neurosurg. Psychiatry* **18,** 44–52.

Milner, B. (1954). Intellectual function of the temporal lobes. *Psychol. Bull.* **51,** 42–62.

Milner, B. (1975). Psychological aspects of focal epilepsy and its neurosurgical management. *In* "Advances in Neurology" (D. P. Purpura, J. K. Penry, and R. D. Walter, Eds.), Vol. 8, p. 299. Raven Press, New York.

Milner, B., and Penfield, W. (1955). The effect of hippocampal lesions on recent memory. *J. Neurol. Neurosurg. Psychiatry* **80,** 42–48.

Milner, B., Branch, C., and Rasmussen, T. (1962). Study of short-term memory after intracarotid injection of sodium amytal. *Trans. Am. Neurol. Assoc.* **87,** 224–226.

Novelly, R. A., Augustine, E. A., Mattson, R. H., Glasser, G. H., Williamson, P. D., Spencer, D. D., and Spencer, S. S. (1984). Selective memory improvement and impairment in temporal lobectomy for epilepsy. *Ann. Neurol.* **13,** 64–67.

Penfield, W., and Milner, B. (1958). Memory deficit produced by bilateral lesions in the hippocampal zone. *Arch. Neurol. Psychiatry* **79,** 475–497.

Phillips, N. A., and McGlone, J. (1995). Grouped data do not tell the whole story: Individual analysis of cognitive change after temporal lobectomy. *J. Clin. Exp. Neuropsychol.* **17**(5), 713–724.

Rayport, M., and Ferguson, S. M. (1996). Psychiatric evaluation for epilepsy surgery. *In* "The Treatment of Epilepsy" (S. D. Shorvon, F. E. Dreifuss, D. F. Fish, and D. G. T. Thomas, Eds.), p. 631. Blackwell, Oxford.

Sawrie, S. D., Martin, R. C., Kuzniecky, R., Faught, E., Morawetz, R., Jamil, F., Viikinsalo, M., and Gilliam, F. (1999). Subjective versus objective memory change after temporal lobe epilepsy surgery. *Neurology* **53,** 1511–1517.

Scoville, W. B., and Milner, B. (1957). Loss of recent memory after bilateral hippocampal lesions. *J. Neurol. Neurosurg. Psychiatry* **20,** 11–21.

Squire, L. (1994). Declarative and nondeclarative memory: Multiple brain systems supporting learning and memory. *In* "Memory Systems" (D. L. Schachter and E. Tulving, Eds.), p. 203. MIT Press, Cambridge.

Wada, J., and Rasmussen, T. (1960). Intracarotid injection of sodium amytal for the lateralization of cerebral speech dominance: Experimental and clinical observations. *J. Neurosurg.* **17,** 226–282.

Wall, C., Wilson, B. A., and Baddeley, A. D. (1994). The extended Rivermead Behavioral Memory Test. *Memory* **2,** 149–166.

Wilkins, A. J. (1986). Remembering to do things in the laboratory and in everyday life. *Acta Neurol. Scand.* **74,** 109–112.

Wilkins, A. J., and Baddeley, A. D. (1978). Remembering to recall in everyday life: An approach to absentmindedness. *In* "Practical Aspects of Memory" (M. M. Gruneberg, P. E. Morris, and R. N. Sykes, Eds.), p. 27. Academic Press, London.

LIFE AFTER SURGERY FOR TEMPOROLIMBIC SEIZURES

Shirley M. Ferguson,* Mark Rayport,[†,‡] and Carolyn A. Schell[§,1]

*Department of Psychiatry, University of Toledo College of Medicine (formerly Medical College of Ohio), Toledo, Ohio 43614, USA
[†]Department of Neurological Surgery, Medical College of Ohio, Toledo, Ohio 43614, USA
[‡]Department of Radiology, Medical College of Ohio, Toledo, Ohio 43614, USA
[§]Medical College of Ohio, Toledo, Ohio 43614, USA

The literature of outcome after temporal lobe excision for intractable seizures is too often restricted to short periods of postoperative observations and/or emphasizes the degree of seizure control without extensive consideration of the details of the patient's life. In fact, a multicenter study of outcome (Spencer *et al.*, 2003) concluded that longer periods of follow-up are essential in order to understand the true effect on the patient's life of eradication of seizures. Our study addressed these two points by presenting long periods of interdisciplinary follow-up, as long as 20.4–24.3 years, with average of 6.6–10.5 years, and detailed, behavioral observations throughout.

I. Introduction

There has been a long path to the present general acknowledgment that achievement of control of seizures is not a sufficient criterion for judging response

[1]Carolyn A. Schell retired in 2003.

INTERNATIONAL REVIEW OF NEUROBIOLOGY, VOL. 76
DOI: 10.1016/S0074-7742(06)76006-8

to temporal lobectomy. Unfortunately, in the name of scientific rigor, evaluation of the effect of surgery on the patient's life brought forth objective methodology in the form of standardized interviews, scales, and so on. Furthermore, behavioral observations were limited, often made just before and at infrequent intervals for short times after surgery. The behavioral discipline, still most often sought by the neurosurgeon, has been neuropsychology, understandable as a reaction to the dread amnestic syndrome that appeared after temporal lobe surgery and was defined by Milner (1954) in the 1950s. In contrast, neuropsychiatry was sparsely represented; surprising, in view of the obvious need, so it would seem, to understand the whole person, as well as the unusual opportunity to work at the brain–behavior interface. This can be attributed partly to the orientation of psychiatry which waited until the 1980s to officially declare an interest in brain-related behavioral symptoms by establishing neuropsychiatry as a subspecialty of psychiatry. However, in the late 1950s, before this recognition of a new subspecialty, one of the authors, Shirley M. Ferguson, M.D., became involved in the behavioral aspects of epilepsy as a result of her unique training opportunities and marriage to an unusual neurosurgeon who was always concerned with the life of his patients, not just their brains and lesions. Some history will shed further light on these developments.

The surgical epilepsy series (1970–1993) of Mark Rayport, M.D., Ph.D., at the Medical College of Ohio (now the Medical University of Ohio) is unusual in its longitudinal prospective nature with study by the same investigators, from initial contact, through the surgical phase, to long-term follow-up to the present, in some cases. The most promise for an ideal outcome resulted from meticulous preoperative definition of the seizure focus and brain stimulation-guided surgery when indicated. This chapter will present data in regard to postoperative outcome in terms of effect on seizures and on facets of the patient's life including his self-evaluation, psychiatric symptoms, marital experiences, and functioning in the spheres of education and work. The neuropsychiatrist carried out Dr. Rayport's dedication by addressing the mechanisms of behavioral problems in terms of brain dysfunction interacting with the patient's life history and psychodynamics.

This contrasts with the more frequent approach of neurological surgery epilepsy centers which emphasizes painstaking definition of the epileptic process and careful surgery, but accumulates behavioral data objectively, mainly from the neuropsychologist, who focuses almost exclusively on cognitive function. If behavioral responses are investigated, the methodology consists of limited interviews and/or questionnaires. Outcome is often reported after short periods of observation.

In this series, behavior was studied by the neuropsychiatrist who maintained observation of and interaction with each patient throughout his course. We point out that support by the epilepsy team makes a significant difference in outcome. Success in accomplishing behavior/life change often requires extensive time periods.

II. Review of the Literature

A little over 100 years ago, Sir Victor Horsley (Horsley, 1886)—one of the two founders of modern neurological surgery—performed his first brain operation at the National Hospital for the Paralyzed and Epileptic, Queen Square, London. This operation was carried out on a patient of the great neurologist, Hughlings Jackson. The patient was a 22-year-old epileptic man whose seizures had appeared after a severe head injury several years earlier. The operation was a success. There were no more fits. How this neurosurgical success affected his life, his person, his functioning is not mentioned. This traditional separation of brain and mind characterizes the orientation of most neurosurgeons even today.

In 1928, Wilder Penfield began his illustrious career in epilepsy surgery in Montreal. With the spread of electroencephalography after 1929, the team at the Montreal Neurological Institute (MNI) directed its attention to the seizures, to localization of the discharging area of the brain, and advised patients with behavioral complaints to seek help elsewhere and return for consideration of seizure surgery after their psyche was mended. In one of a series of classic papers, Rasmussen (1979) reported on Penfield's and his own results. Among 722 patients with intractable partial seizures originating in the temporal lobe, between 40% and 50% obtained complete seizure control. The seizure frequency before and after surgery was the measure of outcome.

Between 1952 and 1972, in the Guy's Maudsley Neurosurgical Unit, London, Falconer (1973) operated on 250 patients with intractable temporal lobe epilepsy (TLE). Many of his patients had behavioral complaints insofar as they were referred from the adjacent psychiatric hospital, the Maudsley. Falconer worked with a succession of psychiatrists: I. P. James (James, 1960), E. A. Serafetinides (Serafetinides, 1962, 1965), and D. C. Taylor (Taylor, 1969, 1972, 1975). These investigators concentrated on the occurrence of aggressive behavior and psychosis and made correlations of these conditions with gross and microscopic neuropathologic findings in the tissue removed at operation. Those patients who showed mesial temporal sclerosis, a lesion occurring at birth and during infancy, were found to have aggressive behavior, whereas a congenital tumor, hamartoma, was associated with psychosis. Although other causal factors were mentioned, such as low IQ, laterality of the focus, institutionalization, educational disruption, medications, and poor genetic endowment, their impact was not directly researched. Taylor (1987) described the assessment of outcome of epilepsy surgery as a "dilemma." He called attention to "the brilliant techniques available to study the epileptic brain and psychometric tools which only partially estimate the predicament of the epileptic person."

Rayport and Ferguson, working at the Albert Einstein College of Medicine during the 1960s, and at the Medical College of Ohio between 1970 and 1993, directed an increasing amount of attention to the "intervening variables" between the epileptogenic processes in the brain and the patient's behavioral condition (Ferguson *et al.*, 1969). With his unusual orientation as a neurosurgeon, Dr. Rayport had a determined dedication to the quality of the patient's brain and behavioral functioning after neurosurgical intervention. At operation, he was in complete control by his meticulous preparation in defining the source of the problem, by being in command of the anatomy and determining what could be done without harm to tissue necessary for vital cerebral function. He always had a clear knowledge of the tissues under his scalpel and carefully sorted out what should and what could be excised. To guarantee that his full attention could be given to the problem at hand, he greatly valued his colleague, J. T. Martin, M. D., a superb neuroanesthesiologist, who safeguarded the patient and informed Dr. Rayport of any changes in the patient's condition which needed to be addressed. To be as sure as possible of the function of the tissues in the field, Dr. Rayport appointed Dr. Ferguson to sit with the patient under the drapes and, in turn, support, interview, and test the patient when the brain was stimulated.

An early expression of this broad interest in the patient's welfare was published by Ferguson and Rayport (1965) in a paper entitled, "The Adjustment to Living without Epilepsy." We found that some patients seizure-free postoperatively had significant behavioral difficulties in their seizure-free life. It was concluded that these difficulties were in the nature of an "identity crisis" for which they required professional help. A few years later, confirmation was published by Horowitz *et al.* (Horowitz and Cohen, 1968; Horowitz *et al.*, 1970). In 1969, we reported on "Similarities in Mental Content of Psychotic States, Spontaneous Seizures, Dreams and Responses to Electrical Brain Stimulation in Patients with Temporal Lobe Epilepsy" (Ferguson *et al.*, 1969). Subsequently, the question of outcome was addressed by defining the patterns of behavioral recovery after successful, seizure-controlling surgery (Ferguson *et al.*, 1993). The effects of surgery on "everyday psychopathology," aggressive behavior, and psychosis were described (Ferguson *et al.*, 1986; Rayport and Ferguson, 1996, 2001). Prognostic criteria based on the interrelations between the individual's psychic life, localization of the seizure focus in the brain, and seizure phenomena were also proposed.

Of course, one must acknowledge the fact that neuropsychological research has made extensive contributions to studies of surgical epilepsy patients. One of the pioneers in this work is Milner (1954) of the MNI, a keen student of cognitive function who made her great contribution, starting in the 1950s, with elucidation of the importance of the hippocampus in memory function. Neuropsychologists working with other surgical epilepsy programs have followed the Milner model by primarily evaluating cognitive function with the goal of determining adequacy of tissue to remain after surgery, cerebral localization of the focus, and prediction of surgical outcome.

Need for a broader evaluation of the person with epilepsy began to appear in the neuropsychological literature in the 1970s. Rausch *et al.* (1977) stated, "Clinicians are becoming more concerned with the postoperative functioning of the whole person." Working with the Minnesota Multiphasic Inventory (MMPI) and a scale developed by D. C. Taylor, Rausch found that good family support correlated with a positive outcome. Of interest to us is that, under her research design, no brain or seizure factors were of prognostic significance. In 1980, Dodrill *et al.* (1980) introduced the Washington Psychosocial Inventory of 132 yes/no items to obtain an objective psychosocial profile of the epilepsy patient. Postoperatively, the test was readministered after an interval of 1 year. Although improvement was shown in test items, there was often a discrepancy with actual functioning. It may be noted that these studies did not separate seizure-free patients from those who had obtained only a reduction of their seizure frequency. Furthermore, in our opinion, the use of a fixed interval of 1 year introduced a significant variable because the time frame of postoperative improvement, in our experience, varied between patients, from days/months to several years.

In March 1987, an article was published in Neurology titled: "Seizure Characteristics, Pathology and Outcome" from the Radcliffe Infirmary (Duncan and Sagar, 1987) at Oxford, United Kingdom. The authors reported outcome after 1 year postsurgery and found that the outcome did not change after the first few years. They concluded that the study confirmed the view that "anterior temporal lobectomy is most likely to have a good outcome if definite pathology is found in the resected temporal lobe and preoperative evaluation aimed at identifying those patients in whom such a lesion is probable." This emphasis on the relation between type of pathology and behavior continues the thesis of Falconer's group at the Maudsley in the 1960s.

Vickrey *et al.* (1992) developed "a health-related quality of life instrument (HRQOL) for patients evaluated for epilepsy surgery." This was subsequently widely used in other surgical epilepsy programs.

In 1996, "Determinants of Health-related Quality of Life after Temporal Lobe Epilepsy Surgery," from the Department of Clinical Neurological Sciences, University Hospital, London, Ontario, Canada was published in Quality of Life Research. In this article, Rose *et al.* (1996) pointed out initially that the influence of preoperative health status and postoperative reductions in seizure activity on postsurgical health-related quality of life are not well understood. They found that among 47 patients who underwent TLX, it was patients with low or medium preoperative HRQOL who had the greatest improvement after surgery. Therefore, they claimed that "the success of epilepsy surgery must be evaluated on the basis of more than just seizure reduction."

In a multicenter study of outcome after epilepsy surgery (Spencer *et al.*, 2003), data included: "seizures, anxiety, depression and quality of life using the standardized measure of the Quality of Life in Epilepsy (QOLIE-89)." "Because the cohort is still being actively followed, we limited analyses of QOL, depression, and anxiety to those observed through 2 years after surgery." They concluded that "true and lasting improvements in QOL are likely to require a lengthy time to be manifest, and are the real goal. Longer term follow-up, in our contemporary, large, and uniformly studied prospective cohort, will be critical in defining the extent and overall impact of epilepsy surgery."

We would respond to such a contemporary study by describing our multidisciplinary work of several decades (Ferguson and Rayport, 1988) in which we emphasized longitudinal multidisciplinary individual case study, beginning with a period of evaluation (usually of 1–2 years) and continuing after surgery as long as possible (from the first operation in 1970, to 1993, at our retirement, when our epilepsy program was officially closed) and, in some cases, to the present, through, mostly, telephone contacts. In this chapter, we present the fate of preoperative seizures with enigmatic shadows sometimes appearing after surgery. We compare the preoperative and postoperative behavioral manifestations, emphasizing self-image, interpersonal relationships, psychiatric symptoms, and ability to function in society. What is life like after surgery for intractable seizures?

III. Material and Methodology

Between 1970 and 1993, Dr. Rayport performed 45 cortical excisions for intractable temporal limbic seizures (TLmS). All patients were studied preoperatively in a longitudinal manner, from 1 month to 20 years (in one unusual case of a patient who was resistant to surgery), with a mean of 3 years. Patients were seen in an interdisciplinary clinic with assessments at each visit by neurology or neurosurgery, neuropsychiatry, neuronursing, and social work. A family member(s) or significant other (SO) was encouraged to participate in developing a clearer picture of the patient's seizures and behavior.

A. Goals of the Preoperative Study

The goals of the preoperative study were to define the epileptic syndrome, localize the epileptogenic zone, establish the intractability of the seizures, and define the behavioral manifestations. The preoperative evaluation and neurodiagnostic procedures which follow, provide the basis to obtain our goal.

B. Presurgical Evaluation and Neurodiagnostic Procedures

1. Confirmation of medication-resistant partial seizures with failure of at least three major antiseizure drugs (ASDs);
2. Neuroclinical history and examination confirming partial seizures;
3. Neuropsychiatric history and examination;
4. Diagnostic electroencephalograms (EEGs) (including T1, nasopharyngeal, sphenoidal, zygomatic, and foramen ovale electrodes) and sleep-deprived studies;
5. Radiological assessment proceeded from skull films, pneumoencephalography, angiography to computed tomography (CT), magnetic resonance imaging (MRI) with temporal lobe protocol, single photon emission computed tomography (SPECT);
6. Stereotactic implantation of depth electrodes for EEG monitoring and brain stimulation;
7. Wada testing for lateralization of language and memory endpoint.

C. Role of the Neuropsychiatric Epileptologist

The neuropsychiatric epileptologist explored the nature and sources of the behavioral problems of the epileptic persons whose complex condition fluctuates over time with occurrence of seizures which may alter cognition, mood, and thought patterns, and may interact with the intrapsychic and the interpersonal life. A case-by-case evaluation was made of life profile, personality, and higher nervous functions and of their interactions with the epileptic pathophysiology. A standardized neuropsychiatric format (Rayport and Ferguson, 1996) was a useful tool for initial evaluation. Observations obtained during office visits and during various neurodiagnostic procedures led to a clearer comprehension of the life of the patient, in terms of understanding the interrelationships between pathophysiologic events of seizure phenomena and brain dysfunction, and psychodynamic and psychosocial factors. Neuropsychological tests were administered as a stabilized battery, optimally at a time remote from the target symptom, the seizure.

D. Demographic Data

Of the 45 patients, 26 underwent left temporal lobectomy for seizure control (TLX) (57.8%) and 19 right TLX (Table I).

Among the 23 male patients, 14 had left TLX and 9 right TLX. Among the 22 female patients, 12 had left TLX and 10 right TLX. Age at surgery ranged

TABLE I
CONSECUTIVE TEMPORAL LOBE SURGERY SERIES

Study population	45 patients	
Male	23	
Duration of seizure disorders (years)	3–52	mean 17.2
Age at operation	8.7–56.9	mean 29.3
Electrophysiological localization of epileptogenic zone by		
Extracranial EEG (EcEEG)	21	46.7%
Stereoelectroencephalography and epidural EEG	24	53.3%
Left temporal lobe surgery	26	57.8%
Dominant temporal lobe surgery	25	55.6%
Duration of postoperative follow-up after TLX (years)	Mean (years)	*Range* (years)
By epilepsy team	6.6	0 to 20.4
By neuropsychiatry	10.5	0 to 24.3

from 8 to 56.9 years, mean 29.3. Among the left TLX patients, 20 were right handed, 4 left handed, and 2 ambidextrous. Among the right TLX patients, 17 were right handed and 2 were left handed. Localization of epileptogenic zone was achieved by stereoelectroencephalography (SEEG) in 24 patients (53.3%), 14 patients with left TLE and 11 with right TLE. Etiological factors were collected historically in the 45 patient as seen in Table II. Neuropathologic findings after TLX in the 45 patients are presented in Table III.

E. Postoperative Evaluation

The entire cohort of 45 patients was analyzed in regard to seizure outcome. Thirty of the seizure-free and essentially "seizure-free" patients were analyzed in regard to behavioral outcome after TLX in terms of self-evaluation, marital status, work, and other activities. This group of patients was chosen because of extensive, reliable data obtained throughout their course from preoperative study to long-term follow-up. These patients were followed by the team from initial contact until closing of the surgical epilepsy program in 1993. There was additional contact in 1995–1996 when patients were recalled for another study. Some of the patients have been followed to the present (Tables I and IV).

TABLE II
ETIOLOGICAL FACTORS OF EPILEPSY IN 45 TLX PATIENTS

Febrile convulsions during infancy/childhood	16
Head injury	15
Brain tumor	7
No known cause	6
Birth injury	5
Brain infection	3
Other brain lesion	1
Alcohol related	1
Preeclampsia with pregnancy	1
Allergic reaction	1

TABLE III
NEUROPATHOLOGIC FINDINGS ($N = 54$)

Neuropathologic findings	N
Mesial temporal sclerosis	29 (63.04%)
Tumor	9 (19.6%)
Cortical dysplasia, heterotopia	5
Gliosis	4
Hypoxic changes	3
Temporal lobe infarct	1
Vascular malformation	1
No pathological diagnosis	1

TABLE IV
FREQUENCY DISTRIBUTION TABLE: POSTOPERATIVE FOLLOW-UP DURATION

Team		Neuropsychiatry	
Years of follow-up	Number of patients	Years of follow-up	Number of patients
0–5.0	20	0–5.0	12
5.1–10	13	5.1–10	10
10.1–15	9	10.1–15	11
15.1–20	2	15.1–20	8
20.1–25	1	20.1–25	4
Total	45	Total	45

TABLE V
POSTOPERATIVE SEIZURE OUTCOME

Postoperative seizure outcome	*N*	Percentage
Seizure-free (SF)	**18**	**40**
Essentially "Seizure-free"	**18**	**40**
Auras only	6	13
One seizure with questionable auras	2	4
Limited number of seizures plus a few questionable unexplained symptoms (auras?) within a circumscribed time period and with a possible precipitating event	4	9
SF with enigmatic symptoms	5	11
Seizure recurrence after a significant SF period	1	2
Continued seizures		
Reduced frequency and severity and/or eradication of major symptom (behavioral) leading to surgery	**3**	**7**
Never seizure-free	**7**	**16**
No data available	**1**	**2**
Reoperations	**5**	**1**
For seizures	3	7
For recurrence of brain tumor	2	4

IV. Results

Although the data are presented in tabular form (Table V), a table is clearly insufficient to capture the nuances and variations of the postoperative course. Therefore, benefits of TLX will be described as follows, not only in terms of seizure control but of other aspects of the patients' lives.

A. Patients Seizure-Free or Essentially "Seizure-Free"

Of the 45 patients, 18 (40%) became *seizure-free* and another 18 (40%) were *essentially "seizure-free."*

Of the 18 (40%) who were *seizure-free*, 3 (7%) had brief periods of seizure recurrence.

KIR was seizure-free for 2 years after left TLX when he had a flurry of auras (10 in 2 months) immediately following brain insult from a motorcycle accident. Simple partial seizure (SPS) and/or complex partial seizure (CPS) recurred at 4.5, 5.5, and 8.5 years after TLX. He had no further seizures for 4 additional years of follow-up. ASDs were never tapered or stopped although he was never reliably compliant and indulged in alcohol periodically.

KOM was seizure-free for 5 years on ASDs and seizure-free 6 months off ASDs. When he developed a paranoid psychosis, he was found to be in limbic status. Psychosis subsided as seizure control was achieved with resumption of a previously successful ASD. He has remained seizure-free for the subsequent 10 years of follow-up (15.7 years after left TLX).

MYD was seizure-free for approximately 11 years after initial right TLX and oligodendroglioma removal. Following reoperation for recurrent seizures and brain tumor, he remained seizure-free for the subsequent follow-up period of 12.9 years. He continues to take one ASD.

Of the 18 (40%) who were *essentially "seizure-free,"* 6 continued to have *auras only*, which did not interfere with daily functioning.

Two were essentially "seizure-free," except for one seizure with questionable auras:

PRS had a single CPS with secondary generalization 3½ years after left TLX. This occurred when she was off ASDs for 3 months and following hyperventilation in response to accidental self-injury. After reinstating one ASD, she has remained seizure-free for the subsequent 10 years. This patient also had auras on two occasions during the first postoperative year and associated with neglect of ASDs, generic carbamazepine (CBZ), and inadequate sleep.

The other, **BRB**, after right TLX, had one CPS, after a 10-year seizure-free period, and during a period of gestational diabetes. This patient also had occasional auras during the first two years, at the time of her menses.

Four were "seizure-free" except for a limited number of seizures, plus a few questionable unexplained symptoms (auras?) within a circumscribed time period and with a possible precipitating event.

COP was seizure-free for 16.8 years after left TLX, except for four CPSs, between 2.6 and 4.3 years postoperative, of which three were nocturnal. All occurred during pyrexia with elevation to 102°F. This patient also had four brief episodes, possible auras, but different from her preoperative pattern: (1) *déjà vu* (at 0.75 years); (2) receptive aphasia (at 2.5 years); and (3–4) alexia (at 10 years). Possible causes for these included neglect of antiepileptic medication and/or stress.

LAK was seizure-free for 3 years after left TLX, except for three seizures at 1 month due to ASD noncompliance. One of two ASDs was discontinued at 0.2 years and the patient stopped the remaining ASD at 0.3 years. Monotherapy ASD treatment was restarted at 0.4 years due to frequent spiking in EEG. Another seizure occurred after about 10 years associated with abrupt cessation of ASD and febrile illness.

LAS, after right TLX, was initially seizure-free on ASDs for 4.3 years except for one seizure flurry at 0.9 years in conjunction with an ear infection and fever of 101°F. Seizures recurred at 4.4 years, 1.1 years after discontinuation of one ASD [sodium valproate (VPA)] and change of CBZ to a slow-release formulation.

After brand name CBZ was reinstated, this patient had no further seizures for 1 year. A probable seizure then occurred after CBZ taper was in progress and the patient neglected to treat a sinus infection; CBZ was resumed after this. He continued to have occasional seizures while newer ASDs were prescribed. Since initiation of a combination of VPA, primidone (PRM), and Phenobarbital (PB), this patient has been seizure-free for 2 years.

NIH, during 18 years of follow-up after right TLX, was seizure-free on two ASDs [CBZ and phenytoin (PHT)] for 4.2 years. PHT was stopped at 3.5 years; CBZ taper was started at 4.2 years. He then had two nocturnal seizures associated with CBZ taper. The CBZ dose was increased; 4 months later he had five nocturnal seizures over a 3-month period and CBZ dose was increased again. Thereafter, he had an unsatisfactory response to newer ASDs as manifested by continuing, although infrequent, seizures. Once he was returned to his original preoperative ASD regimen, he became seizure-free. Most recently, he has been seizure-free for 7 years on two ASDs.

Five patients were seizure-free except for enigmatic events (probably not seizures).

SMA was seizure-free for 10 years after left TLX, except for one enigmatic episode after 3.8 years: This was more likely a hypoglycemic event because of the pattern of the episode and its association with prolonged food deprivation, emotional trauma, and a new over-the-counter (OTC) drug. This seizure occurred after patient had been off ASD for 0.8 years. After the episode, the ASD was reinstated and the patient was seizure-free on one ASD for 6 more years of observation.

CAK was seizure-free after right TLX for 21.4 years on one ASD, PB, continued at patient's request. She would have questionable infrequent "auras," usually experienced just prior to her yearly follow-up.

PAB, after right TLX, was seizure-free on one ASD for 24.3 years except for auras and/or nonepileptic seizures associated with various types of stress: emotional, environmental heat, and pyrexia. There was also manipulation of ASDs by patient. Four generalized tonic clonic (GTC) seizures occurred during the first 2 years and two CPSs at 10.8 years with reduction of PHT dose.

RUD was seizure-free, after right TLX, for 19 years on one ASD except for one GTC and CPS at 0.7 and 6.3 years, respectively, after sudden suspension of ASD. Three dissociative episodes occurred between 5 and 7 years after TLX in response to reduction or discontinuation of ASD and frustrating personal situation with excessive work load (60 h/week), conflict with peers, reduced income, and divorce.

LAK was followed for a total of 6 years after left TLX. She was seizure-free initially for 2.2 years on ASDs until she independently stopped one ASD at 2.2 years. She was then seizure-free for 3.8 years off ASDs. At last contact, diagnosis was in question as to dissociative disorder versus seizure. She was subsequently lost to follow-up.

One (2%), **SHR**, after left TLX, had *seizure recurrence after a seizure-free period of 3.4 years when he stopped his ASDs independently.* His seizures recurred and worsened even though ASDs were restarted: We were later informed that he underwent reoperation elsewhere for intractable seizures.

B. Continued Seizures

Three continued to have seizures after left TLX but benefited from reduction in frequency and severity. *Two of the three obtained improvement after reoperation.*

JAJ was seizure-free and behaviorally improved for 1.8 years after left anterior TLX. Seizures and difficult behavior recurred despite continuation of ASDs. On reopening of craniotomy, 3.5 years after first surgery, extension of hippocampectomy was not possible. One year later, brachytherapy was utilized with I-125 into the remaining hippocampus. This patient was followed for an additional 2.8 years during which improvement was noted with reduction in frequency and severity of seizures and elimination of severe aggressive behavior, which had initiated the surgical treatment. Twelve years after the brachytherapy, aggressive behavior recurred in association with partial seizures related to ASD changes. Neuropsychiatric evaluation discovered that ongoing interpersonal dynamics duplicated his rejection by his original family. Total duration of time patient seen after first surgery: 16.2 years.

PEL was seizure-free for 1 year after her first surgery for brain tumor and seizures localized to the right anterior temporal lobe. Tumor recurrence led to a second surgery, 5 years after the first surgery. She was never again seizure-free although seizures were reduced in frequency and severity which enabled her to live independently. Total duration of follow-up after second surgery: 7.8 years.

C. Never Seizure-Free

Seven (16%) patients were never seizure-free after TLX. Two of the seven underwent reoperation but did not gain seizure control.

D. Reoperation

Five (11%) underwent reoperation.

Of the three (7%) reoperated for seizures, none became seizure-free, but **JAJ** had significant improvement in behavior (aggression) as seizure frequency was reduced. A seizure mechanism had been clearly established as the precipitant for the behavior disturbance.

Two (4%) were reoperated for recurrence of brain tumor. **MYD** became seizure-free (as described earlier under the seizure-free category).

E. Challenges to Evaluation

It is of interest that although these patients had been considered seizure-free postoperatively, careful review of the clinical notes reveals that, in eight patients, symptomatology appeared which was attributed by the patient to seizures which, on closer scrutiny, had another explanation.

Sometimes, a familiar, clear-cut isolated seizure was described. More often, sensations not definitely denoting a seizure were reported. It was important to study these occurrences in the ongoing context of ASD intake and the dynamics of the patient's life.

CAK, followed for 21.4 years, would note "an unpleasant taste," (her preoperative aura) or brief light headedness just prior to her yearly visit.

COP had equivocal symptoms, often indefinite, such as "headache," followed by a "dreamy feeling" and lasting for 1 day. This occurred during tapering of medication.

BRB (off medicines at 4 years postoperatively), reported a 30 second "aura" at the time of her menses, 8 years after operation. Ten years after operation, when she was 14 weeks pregnant, she experienced a chewing episode (at the dentist) and a nocturnal seizure observed by her husband. Later, she was found to have gestational diabetes and lost the child. No further episodes occurred during the subsequent 8.8 years of follow-up.

MYD experienced a generalized seizure 8 years after surgery. He had been off PRM for 6 months, but was still on CBZ. He stated that he had disrupted his schedule by sleeping in and omitting a dose of CBZ. Subsequently (12.9 years), he has been compliant and has been seizure-free.

SHB reported continuing epigastric auras described as "chest pain." There was the possibility that she was identifying with the symptoms of her mother who became a cardiac invalid and required constant companionship (provided by the patient until mother's death, 6 years after patient's operation).

RUD had a GTC seizure on the first postoperative day due to erroneous ASD withdrawal. One year after TLX, he had a GTC seizure after omitting evening and morning doses of ASD and in a sleep-deprived state. Five years after surgery, he reported an amnestic period whose differential diagnosis of seizure with amnesia or a dissociative episode was unclear. A dissociative etiology was strongly considered because the period of amnesia seemed to have strong triggers of inappropriate remarks at work, threat of dismissal from his job, and ambivalent feelings in regard to the work setting, prolonged work hours, absence from family, and so on. A genuine seizure was entertained because of the symptoms of abdominal sensation and loss of a period of time plus aggravating features of recent inadequate rest and diversion. At 105 months after surgery, the team neurologist considered that RUD had had episodes of transient global amnesia when he was under emotional stress.

V. Behavioral Outcome

The remainder of this chapter presents the behavioral outcome in the 30 patients in the seizure-free and essentially "seizure-free" categories in terms of self-evaluation, marital status, work, and other activities.

A. After Left Temporal Lobe Excision

1. *Self-Report*

All but one of the patients who underwent TLX spontaneously reported a positive change in their self-evaluation, mood, and interpersonal relating. The changes occurred during the first year after operation, endured throughout follow-up (with some fluctuation in response to life events), and were expressed as:

Change in way I feel about myself
I stick up for myself
I have a new assertiveness
I can look people in the eye
Things that used to bother me don't
Best ever felt in my life
Dysphoria gone
Guilt is less
I'm like a freed man
More at ease
More full of life
Positive mood despite problems
Pleased with ability to take initiative and problem solve
Arousal of sexual feeling (reported by two patients)

2. *Fate of Preoperative Psychiatric Symptoms*

KOM had always been a loner with a paranoid tendency. He developed an acute paranoid psychotic episode 5 seizure-free years after TLX, 6 months after physician initiated tapering and discontinuation of his ASD and in response to recurrence of seizures with temporolimbic status. Subsequently (over 10 years), he was seizure-free and remained a loner with easily aroused paranoid ideation, but without psychosis. He was helped by haloperidol, started at the time of the psychotic episode, ASD, and supportive therapy.

FRS, a teenager, had an unresolved grief reaction in response to her mother's death prior to surgery. This appeared to be due to the activation in her seizures of vivid visual memories of experiences shared with her mother. After surgery, over months, there was spontaneous resolution of the grief reaction, as patient was relieved of the constant vivid reminder and was able to face and cope with her loss.

KIJ described himself as a life long "worrier." After surgery, he showed some obsessional thinking about his diagnosis when the pathology report suggested the possibility of a tumor (even though promptly refuted) and in response to back pain which he identified with his uncle's complaint, caused by an aneurysm. He had no clinically significant obsessive compulsive symptomatology.

Two women were concerned about recurrence and were especially wary at the time of their menses. As they remained seizure-free, the concern faded (in 9 months in one and in 25 months in the other).

3. *Transient Postoperative Depressive Symptoms*

FRS, as mentioned earlier, suffered a continuation of her preoperative grief reaction which spontaneously and naturally resolved within the first year after surgery.

COP, married and mother of two daughters, experienced depression in response to failure of family to redefine their way of relating to her changed, nonepileptic persona. Psychotherapy with patient and family brought relief.

WEN, a divorcee prior to surgery, had an adjustment reaction precipitated by an incapacitating fracture of her leg and loss of a boyfriend.

LAK, a single mother of two following divorce (prior to surgery), was greatly disappointed by the occurrence of a seizure precipitated by a febrile illness when she had abruptly discontinued ASD because of financial straits. She was also struggling with practical life problems. Depressive feelings rapidly resolved when she recovered from her illness, resumed ASD, and was reassured and supported by the team. She remained seizure-free over the next 7 years.

BASA, a married mother of one child, experienced depressive symptoms with guilt when it was necessary to place her mother in a nursing home because she was showing psychiatric symptomatology. Her mother had always been critical and rejecting in response to her daughter's epilepsy. At that time also, her adolescent daughter was becoming more demanding and critical. Psychotherapy was helpful in resolving her symptoms.

KIT, a 54-year-old married father of two, developed anxiety symptoms, which were first interpreted by him as seizures, in reaction to the stressful situation of his independent business. Careful neurological reevaluation, psychotherapy, and marital counseling led to disappearance of this reaction. He has been seizure-free for 10 years following surgery.

4. *Marital Status*

Three women and four men had married prior to surgery. Two women had children. **BASA** had one child and **COP** had two. **SMA** had been married for 10 years but never fulfilled her hope to have a child. Postoperatively, as the couple was trying to arrange for an adoption, she became pregnant and delivered a normal child at term.

Two young women, in high school at the time of surgery, married after operation. **CRT** had two children. The other, **FRS**, was disappointed that she did not become pregnant and when she sought specialist advice, was told that she had polycystic ovaries.

Four women had been divorced prior to surgery. Two remarried happily after surgery, but became widows: **PRS** after several months and **ARM** after 2 years. Both resumed employment and remained independent.

KOM, the loner described earlier, remained single.

5. *Activities*

Comparison of functioning before and after TLX showed a palpable improvement in all. Those who were working preoperatively made significant strides as the quality of their work after surgery demonstrated. Time parameters varied, but the majority of the patients took their first emancipatory steps within the first year after operation. Exploration, unsuccessful attempts at study and training accounted for postponement, in some, of their first job to months or years after surgery. There was an impressive determination on the part of these individuals to take advantage of the new life they felt could open for them, in view of their freedom from seizures and changed, positive self-image, as described earlier.

Before surgery, three patients were employed; five were unemployed, two were in high school; one did volunteer work; three were homebound.

a. Employed. Two women and one man were employed.

BASA was an elementary school teacher who had had difficulty both in association with seizures (aphasic and amnestic manifestations) and ongoing cognitive slowing (difficulty learning names of pupils, performing her tasks). After surgery, she had no interruptions, could perform tasks more quickly (checking papers), and knew children's names by end of first day of school. Gradually, she was given assignments to higher grades (initially teaching second grade, she was advanced to third and, then, fifth). Despite hesitation in taking on the more challenging task, she proceeded and did well. She commented that, "I couldn't have taught such a class the way I was."

KIJ had worked in a factory as a supervisor. The independent aspect of his work had allowed him to cope with his work despite the seizures. After surgery, he returned to his work site and was given promotions.

WEN had been employed in a semi-sheltered position in a welfare department office. Prior to surgery, she noted fluctuations in her mood in the preictal state: Sometimes, she would be in a negative mood, feeling that everything was wrong, that she could do nothing right. At other times, she would became very talkative, elated, and make inappropriate advances to male workers. Behavior was never prolonged enough to cause termination of her work. After TLX, mood was stable, and, over time, she not only maintained her employment, but was given promotions.

b. Unemployed. Four men and one woman were unemployed prior to surgery.

SWC, 39 at operation, had worked as a draftsman but had been unemployed for 4 years because of the seizures. Postoperatively, he returned to school but did not succeed in studies involving verbal ability. He persisted and enrolled in a maintenance course for which he showed aptitude. He acquired a supervisory maintenance job in an upper class hotel and worked happily and continuously for 4 years until a back problem interfered.

DRO, 22 at TLX, had not been employed for several years. After surgery, he found a job at a golf course in another state. He was, then, lost to follow-up.

KOM, prior to surgery, had been employed in a greenhouse until he relinquished the job when work-up for possible surgery was initiated. After surgery, he remained at home with his parents, communicating little, because of sensitivity to a paranoid degree. He was seizure-free on ASD for 4 years and off medication for 2 years when he developed a florid paranoid psychosis which was associated with temporolimbic status. Once the seizures were brought under control with resumption of ASD, the psychosis subsided. Haloperidol, also prescribed, was continued and was of some help in ameliorating his paranoid thinking. He greatly resisted attempts at rehabilitation, being supported by disability funds and his family. Finally, after several years, he decided to pursue his interest in wood working and planned, without the parent's knowledge, to move to another state where he spent a year at a specialized school. Unfortunately, he was not successful and returned home. Once disability funds were terminated (because of his seizure-free state), he agreed to look for work and about 10 years after surgery became a full-time maintenance employee in a department store where he chose the night shift which limited his need for interpersonal contact. When last seen, he had been employed for 5 years.

KIT, 54 at the time of surgery, had had four jobs, each lasting 4–12 years, until 4 years before surgery. He then was able to work, only briefly, as a security guard 2 years prior to surgery. After surgery, he took a course in small business management and conducted a concession in the police department, until he decided to retire when factors beyond his control induced an anxiety reaction. Subsequently, he was happily engaged in volunteer work in a hospital.

PRS, before surgery at 30, had had short-lasting (weeks to 2–3 months) jobs which would be terminated because of the occurrence of seizures. For 6 years prior to surgery, she was unemployed. After TLX, she was steadily employed in a deli department of a large grocery store.

c. Education. Prior to surgery, the performance of two high school students had dropped.

CRT, 15 at operation, described the toll that the seizures had taken on her abilities. Seizures would cause her to lose continuity with activity in the classroom. When she "came to," she had lost her place and, sometimes, would be mocked for the strange behavior or comments she had made during the seizure. She was largely rejected by her classmates and chided by her teacher for resting her head on the desk when the sedative effect of the ASD took over. She became irritable and depressed. Postoperatively, her school performance improved impressively and peer relationships were normalized. After graduation from high school, she, very early, took two jobs because of her desire to make money for tuition at college. At first, she had basic jobs, such as waitress in a fast food restaurant, but in a short time, she became an assistant manager of a photo department. She abandoned her plans for more education when she married. She had two children and helped her husband at home in his business. She was exceedingly grateful for the conquest of her epilepsy and ecstatic about her life as wife and mother.

FRS, 14 at operation, had been doing poorly in her high school studies. She was depressed and irritable, often striking out at her peers. Her seizures contained components which accentuated the sadness of her mother's fatal illness and, finally, death.

She had become the companion of her mother who had terminal cancer. During the aura of her seizures, she would have a visual aura in which she would see a program which she had watched with her mother. After her mother's death (prior to surgery), her seizures provided constant reminders of her lost relationship and shared experiences. Following surgery, in her seizure-free state, she was able to spontaneously resolve her grief reaction over several months, converting the acute pain to memories which she cherished. Initially, she had planned to go into nursing, but gradually realized that "the science things" were an obstacle. Helped by sensitive counseling, she decided to switch to another path, and took courses in preschool education. Since completing the course, she has been employed steadily in a child care setting. She is happily married, but regrets her inability to become pregnant. She has been told by an infertility specialist that she has polycystic ovaries.

d. Volunteer Work. **ARM**, 29 years at operation, had only been able to do volunteer work. Twelve months after surgery, she completed a business school course and 2 years after surgery found employment at a department store. She kept her job until her husband's illness required her to remain at home for 2 years. After his death, she returned to the work force and had remained steadily employed for 5 years when last seen.

e. Homebound. Three patients were homebound.

COP, 39 at operation, did not dare venture out often. She spent her time maintaining her household, raising her two daughters, and pursuing a knitting project at home. After surgery, she was determined to lead a more normal life, hopefully finding a job which would utilize her arithmetic skills. At 4.5 months, she arranged for volunteer work in a hospital and, at 18 months, took a job as a bookkeeper, enrolled in a course, and obtained a real estate license. At 31 months, she had a seizure and hesitated to continue these activities. At 43 months, she again ventured out and found work in the bookkeeping department of a fast food chain. Subsequently, she had various jobs, mostly temporary. It was thought that her difficulty in finding full-time work might be explained by her not being computer savvy and refusing to take a course because she thought her memory function was not adequate. Objective memory testing did not confirm significant impairment, but her hesitation could not be overcome. She continued to insist on maintaining her independence, refusing offer of Bureau of Vocational Rehabilitation help and finding various short-lasting opportunities.

LAK, 25 at operation, had separated from her husband prior to surgery and was caring for their two young children. About 4 months after TLX, she found a job in her favorite area of food preparation (which she had emphasized in high school) in the dietary department of a nursing home.

SMA, 29 at operation, was married and had been confined to home under constant supervision because of seizure events dangerous to her or others: If in the kitchen when a seizure occurred, she might put her hand on a hot burner. If someone was around with whom there had been negative feeling, she might physically attack that person. After surgery, she was freed. She took over her household. In about 1 year, she became involved in door-to-door proselytizing for her church and, at 4 years postoperation, was able to be a child care giver for three to five children. After many years of marriage and inability to conceive (and being told that she could never have a child), she became pregnant and delivered a normal child at term. This was the high point of her life. Her husband agreed, commenting: "That's what she needed."

B. After Right Temporal Lobe Excision

1. *Self-Report*

All 10 patients who underwent right TLX expressed very positive feelings about their self-evaluation, beginning in the first months after surgery and continuing throughout the follow-up.

Their spontaneous comments included:

I'm so grateful
Mood is real good
I have a wonderful sense of release
I have more energy
I'm joyful, very relaxed
I'm enjoying life to the fullest
I have decreased seriousness—a new joking around
I used to be mad and blew up. Now, I sit down and think
I'm more confident
I'm less easily intimidated
I don't have so many fluctuations in thinking
I don't let things get me down
I'm no longer an epileptic. I'm not half a person

2. *Short-Lasting Depressive Symptoms*

BRB had a postpartum depression. (The baby was born 5 years after surgery. Pregnancy was postponed until ASD was discontinued.) Patient had a traumatic delivery. The baby suffered a forceps injury to his head (with no immediate apparent consequences). The patient attempted to nurse her baby, but the baby did not thrive and required brief hospitalization. She felt that she had failed. In addition, she began to evaluate her husband and concluded that she had made a questionable choice, a compromise deemed wise at the time, because of her disability. Once the neonatal period became more stable, and she engaged in psychotherapy, she was able to gradually emphasize her husband's positive traits, tap into her warm feelings for him, and achieve a satisfactory relationship.

RUD married 1 year after operation, fathered two children, but then became more and more dissatisfied as his wife's immaturity became manifest. After a marriage of 6 years, he was divorced. Two years later, he remarried most happily. His wife was a social worker at the nursing home where he was working as a licensed practical nurse (LPN). Still attentive to his biological children, he and his wife devoted themselves to caring for physically and/or emotionally disturbed children, first as foster, and, then, as adoptive parents. They assumed parenting for four children. Tragedy occurred 13 years after marriage when his wife died of a cerebral aneurysm. Depressive feelings were understandable. **RUD** coped remarkably, continuing his work and finding ways to care for the children.

PAB experienced a depressive period in response to changes in life situation: She was adjusting to the decision that she was not able to join the work force and to her landlady's move which separated her from means of independent transportation. Time and supportive help from the team brought resolution.

3. *Focus on Possibility of Seizure Recurrence*

CAK overinterpreted sensations: If a "twitch" occurred, she thought she had had a seizure. When under pressure, she might feel "lightheaded." Even after many years had passed since surgery, she would report a "funny taste" just prior to her annual visit. She also stated her preference to remain on some ASD (PB). She was given reassurance, recognition for her advances in socialization, and promised availability of the seizure team which would continue monitoring all the surgical patients whether seizure-free or not.

PAB would report possible seizure-related symptoms which were questionable in view of the context of their occurrence and her asking for reassurance that we would continue to see her. Once she remarried and felt more secure, these complaints no longer were mentioned.

4. *Marital Status*

Four female patients had been married before surgery.

PAB had been divorced from a husband who resented her disability and did not provide for her medication. (In fact, her mother gave her medication as a wedding present.) After she produced two children, her husband arranged for her institutionalization in a state psychiatric hospital, where she remained for 5 years until a physician, who had known the patient and family when he worked at that hospital, arranged for her to be transferred to the state hospital in our city so that she could be studied for possible surgery for her epilepsy. Thirteen years after surgery, she remarried happily.

BRB had a period of depression when she evaluated the husband she had chosen against who might have become her life partner. With help, she emphasized his positive qualities, excused his gaucheness, remained with him, and was happy to become the mother of his children.

COR was finally abandoned several years after surgery by her husband from whom she had previously (presurgery) separated and divorced and later, resumed cohabitation, although she would not remarry him. They had always had a stormy relationship, accentuated by her paranoid personality and opposing ideas about child rearing.

CAK was married, prior to surgery, to a man who had personality problems, came under legal surveillance, and was lacking in attentiveness. She chose to stay in the marriage for security and for the sake of their two children.

Two men married happily for the first time: **BAS**, 12 years after surgery at the age of 33 and **NIH**, 8 years after surgery at the age of 47.

MYD was pleased with his SO arrangement beginning 6 years postoperation.

RUD, married twice after surgery. (1) One year after operation he married a woman who proved exceedingly immature. He conscientiously shared responsibility for their two children after the divorce. (2) Eight years after operation, the patient, then an LPN, married a woman who was a social worker at his place of employment. They were very devoted to each other and decided to contribute to the care of disabled children, first as foster parents, and then adopting four children. They were very satisfied with their relationship and dedication until the wife's sudden, unfortunate death from a missed cerebral aneurysm about 13 years after marriage and 21 years after surgery.

5. *Activities*

a. Employed. Four patients had been steadily employed prior to surgery. After surgery, two patients happily returned to their places of employment. In each case, there was gradual advancement and/emancipation.

BRJ had worked as a bagger in a grocery store. After surgery, he returned to his job and, after a short time, was assigned to various areas—"wherever they needed help." His work week increased from 40 to 55 h. In addition, he worked on weekends as a farm hand for eight farmers.

MYD returned to the company where he had originally obtained a job (with the help of his father, a longstanding employee), doing electrical work. He returned to this job, but, after a time, decided to sever the tie and find work elsewhere. In the new location, he was very successful, earning promotions, writing training manuals in the course of instructing others, and began to think of starting his own business.

Two patients returned to their previous activity, which they described as "boring" or confining and, shortly, dared to leave the secure, sheltered setting:

BRB, a married woman, took the initiative to leave the secure setting of the family business where she had been under constant protective surveillance for several years. She enjoyed being a full-time homemaker and, also, addressing her own interests. She studied music, practicing the piano 3 h a day, took up the guitar, and played for church services. Eventually, she took a course in medical assistance and finished at the time of the birth of her child, 5 years after surgery (pregnancy postponed until she was off ASD). She postponed looking for work, but did maintain outside activity by joining a semiprofessional singing group.

BAS returned to stockroom work which he found "boring." He enrolled in college courses, but was unsuccessful. He sought vocational advisement, not immediately helpful, and, after a considerable time and exploration, combined his interest in photography and intrigue with the hospital setting to enroll in a course for X-ray technology. After completing the course in 2 years, he found employment, first in a doctor's office, and, then, in a hospital and has been working steadily for 12 years beginning 7 years post-TLX.

Two male patients were faced with finding suitable employment after living under the limiting burden of their illness:

NIH, 39 at surgery, had emigrated from the Middle East at the age of 19. He had obtained a B.A. degree over time, but never achieved steady employment. He returned to college, taking graduate courses in public health. Progress was slow as he would enroll in courses, be impeded by interruptions from his seizures and their aftermath, and, then, drop the course, only to start again the next semester. Postoperatively, he returned to his only familiar activity, of study, but gradually realized that he should cease being an eternal student. Considerable time passed until he was referred by the team to a very skilled vocational counselor who helped him to find a suitable job, as a laboratory technician. He was employed in a private company for 2 years, and then, obtained a job in a college research laboratory. He remained there for 6 years, until the grant supporting his salary was not renewed. After a hiatus, he found work as a substitute teacher in the public school system. He greatly enjoyed the work, was very successful and, finally achieved a full-time position. He is planning on getting a teaching certificate, now, 18 years after operation.

RUD, 24 at surgery, had been occupied preoperatively by doing odd jobs. After surgery, he spent considerable time attempting studies which would prepare him for work in conservation. When that did not materialize, he decided to enter a course to become an LPN. He succeeded and has been employed in a nursing home for 13 years.

b. Unemployed. Two women had been impeded by behavioral concomitants of their seizure disorder. Postoperatively, there was significant improvement in their functioning.

PAB, a 27-year-old divorced mother of two sons, had experienced a lifelong struggle with seizures and their behavioral concomitants in a rejecting primary family and, later, with an unsympathetic, abusive husband with whom she bore two children. Her husband arranged for her admission to a psychiatric hospital (where she remained for 6 years) because of the paranoid thinking and aggressive behavior later determined to be clearly related to her seizures. After surgery, she was gradually released to independent living via a group home and the YWCA. She has had no psychiatric episodes or hospitalization since TLX. She was motivated to work but was unable to maintain adequate speed for simple assignments. After 2 years of trying to maintain a job (e.g., in a restaurant), it was decided, with input from a counselor, that she could not enter the work force. She became a volunteer at an epilepsy center, wrote articles (published) about epilepsy and poems, and maintained a satisfactory daily schedule. She recognized the fact that she was rejection sensitive, that she could carry on a one-to-one relationship, but felt uncomfortable in a group. She was greatly supported by a vocational counselor, who maintained regular contact and introduced her to and taught her home-based skills of shopping, cooking, sewing (made clothes), and knitting. In this context, the counselor became the friend and caring parent figure she had never had. Thirteen years after surgery, she remarried. When her husband developed a kidney problem and required dialysis, she maintained his schedule of treatment appointments, an appropriate diet, and so on, and appreciated the companionship.

COR, a 36-year-old married mother of three at the time of surgery, had repeated negative experiences in all circumstances of school, work, and so on in response to her paranoid personality, later seen to be associated with frequent limbic system status. After surgery, she had some success at employment, some amelioration of her reactivity, but the reactions of her dysfunctional family interfered with what seemed the beginning of a healing process and old patterns of reacting to trying situations recurred. She continued to strive for independence even when deserted by her family.

c. Homebound. Two women were homebound prior to surgery:

CAK, a 27-year-old married woman and mother of two children at surgery, was fearful of going out, feeling shame if she had a seizure. [A seizure had become the symbol of the out of wedlock pregnancy (baby given up for adoption), during which she suffered preeclampsia, cause of her first seizure.] After TLX, she sought out activities in the community: She worked in her husband's office, joined the Welcome Wagon, participated in activities at her children's schools, and so on.

SHB, a 24-year-old single woman at the time of operation, had been expelled from school (with no arrangement for further education) at 13 because of aggressive behavior associated with her seizures. She had coped with her isolation and lack of structured activity by reversing day and night. Her main activity was making wire sculptures. After operation, challenged by her ill mother (a cardiac invalid), she not only adopted a normal diurnal/nocturnal schedule, but became mother's steady companion and assumed housekeeping tasks. After her mother's death, 6 years after surgery, she continued to maintain the home for her father. When last seen 20 years after surgery, she had satisfactorily completed a work evaluation at Goodwill and with guidance by an excellent Bureau of Vocational Rehabilitation (BVR) counselor, she had been employed in the dietary section of two nursing homes, one for 8 months and the other for 1 year. She then worked in the deli department of three supermarkets for a total of about 6 months. After her father's death 20 years postoperatively, she obtained a driver's license, became active at church and for the first time in her life, made friends outside the family.

VI. Discussion and Conclusions

Although outcome studies have appeared in the literature for decades regarding outcome of temporal lobe surgery, basic questions must still be asked. If patients are reported to be seizure-free, at what point is that determined?

In our patients, three had recurrence of seizures after being seizure-free for several years. Recurrence was associated with events which preoperatively had proven etiological for seizures, such as high fever, infection, and/or changes in medication. In regard to ASDs, it appeared that as much study and care were required during follow-up as in the preoperative evaluation period. An outstanding finding was the frequent failure of the newer ASDs to provide control. As in depression, we found that if a drug or drugs had been associated with seizure control, it was wise to reinstate that drug or combination, rather than taking the patient into new territory. Not only did the newer ASDs not control seizures, but caused significant negative side effects. Furthermore, delay in reinstating the previously successful postoperative ASD(s) seemed to provide a new situation in which control was harder to achieve. However, once the original ASD regimen was reinstated control was possible. It should be noted that during the seizure-free period, two patients, **LAS** and **NIH**, were able to make great advances in their lives:

LAS attended, and performed with honors, at a community college in a course addressing computerized tool and design. He was then able to return to his previous employment site (tool and dye worker) from which he had been discharged because of his seizures 4 years before his operation. On returning, he assumed a special role, being the only person in the company specially trained.

NIH, an eternal college student, became a laboratory technician, employed at one private company for 2 years and, then, became a research lab technician in a college setting. He was employed for 6 years until grant money was no longer available. Fortunately, recurrent seizures were nocturnal so that employment was not interrupted. Understandably, he was very concerned that he might have a diurnal seizure which, most likely, would have jeopardized his work. After leaving the laboratory, he sought employment persistently until he decided to apply for work as a substitute teacher in an elementary school setting. He accepted assignments to grades one through six and became so successful that he was regularly assigned and is now employed full time. He reports that, "I never missed a day." He is happily married. His wife, a nurse, still works in the hospital setting and their 7-year-old son is very bright and advanced, speaking four languages, under the instruction of his father. **NIH** plans to acquire a teacher's certificate. Recently contacted, 19 years after surgery, he has had no seizures for 3 years, since his original medications (PHT and CBZ) were reinstated, and his ASD schedule customized: In view of the mainly nocturnal occurrence of the seizures, he takes medication three times a day with a larger dose at bedtime.

Another challenge is to accurately define the symptoms reported as seizures and also attend to stress-related reactions. Could episodes be nonepileptic seizures, perhaps anxiety attacks, which may have origin in the current life story? Some "seizures" were suspect, specifically those that occurred diurnally and in dynamically meaningful situations. One example will suffice: After **NIH** underwent TLX, he remained on PHT and CBZ and was seizure-free for 3 years during tapering of his medication. A new outside neurologist supervising his medication introduced a new ASD. Seizures recurred though infrequently. When contacted 9 years after TLX, he gave a history of his recurrent seizures, occurring at first 2 years apart, then, twice a year, and at that time, three in 6 months. The content of the episodes and the context of their occurrence suggested that he might be experiencing nonepileptic seizures. Eight years after surgery, he had married a nurse whom he met on a visit to his native home. At the beginning, there were sexual difficulties which required gynecological intervention.

The first "seizure" occurred 1 month after his marriage and the second, 2 months after marriage. Both were nocturnal. His wife described the event: "I'm panting, my arms are bent and I'm trying to shake." On another occasion, when he took his wife to the emergency department (ED) because of chest pain, he experienced symptoms while in the waiting room: "I almost fell down. They seated me." He was told that he had been overbreathing.

NIH described his dream life which he believed related to his seizures. The content of his dreams involves his father who died 8 years back. Patient states that he was greatly influenced by his father, who would speak of doing the right thing. He taught him to "live a good life, to be cooperative and honest. He told me a lot of things—how to care, not to ignore." Patient recalls that when he had the episode in the emergency department, he presumed it was a seizure. "I didn't want anyone to think I wasn't taking care of my wife." Other concerns at that time included reduction in sexual feeling and change in his job, with reduction to half time.

After a long period of trial of newer ASDs, his original prescription of PHT and CBZ was reinstated, with attention to timing of doses, in view of the nocturnal occurrence of the episodes (that were most likely genuine seizures). He has now been seizure-free for 3 years.

There was also the nagging question as to whether he might have had recurrence of the small tumor found at the time of surgery. MRI was negative.

When under stress, he also again manifested mild obsessional thinking. When this became intrusive, paroxetine was prescribed. He found this helpful and continues to take it along with his ASD.

LAS had a similar experience related to medication: Ten months after TLX, he had a flurry of seizures in conjunction with an ear infection accompanied by fever of 101°F. He was seizure-free for 4 years when he decreased his ASD on his own and suffered a nocturnal seizure. He too was tried on newer ASDs (prescribed by an outside neurologist) for a considerable time during which he had some seizures, none adversely affecting his excellent job as a computerized tool and dye worker. Finally, since returning to a combination of his original ASDs (VPA, PRM, and PB), he has remained seizure-free.

KOM was seizure-free on ASD for 6 years and off ASD for 2 years. He developed a florid paranoid psychosis which was found to coincide with the occurrence of temporolimbic status and cleared when seizures were brought under control. No further psychotic symptoms were manifested during the subsequent 10 years.

Over and over, patients would thank us for helping to make a change in their lives. They appreciated the consistent supervision and follow-up by doctors whom they well knew and who were readily available for the specific problem at hand. They appreciated the input of the team who provided analysis of their difficulty, supervised adjustment of their medication, explained and reassured, supported guided introspection, helped with family issues, and referred them to sources skilled for encouraging and guiding them to suitable study or work. The benefit of skilled epileptologic specialists was evident.

Almost all patients expressed remarkable changes in how they felt about themselves and how they were able to engage in satisfying interpersonal relationships after prolonged estrangement from normal life because of their seizures. It was as if the monkey had been taken off their backs. Some, so inexperienced, found it difficult to form relationships and welcomed help, in group therapy, or individually, such as working with a counselor who invested time in making the patient independent and productive in the home setting. Some came to accept their social limitations and were satisfied to function in whatever circle might be realistically defined.

As we previously reported in the literature (Ferguson and Rayport, 1988), the behavioral problems of our patients prior to surgery could be due to extrinsic factors, such as rejection by family and neighbors, or in social settings, at school or work, because of the fact or manifestations of their epilepsy. In some, the behavioral symptoms were an intrinsic part of their epilepsy, with the appearance of mood change, psychotic symptoms, or aggressive behavior. Once seizures were controlled, psychiatric symptoms, psychiatric hospitalization, and serious interpersonal friction no longer occurred, unless seizures reappeared.

Success in work was the gratifying bonus of the seizure-controlling surgery. Patients who had been employed returned to work and were promoted. Others, who had not worked or had been unemployed for several years, obtained training and/or entered the work world at a gratifying level such as computer-guided tool and dye worker, research lab technician, X-ray technician, LPN, teacher, skilled maintenance worker, or child care teacher. Vocational choice often reflected their positive experience during their treatment. Pursuing the dedication to the longitudinal follow-up, we have been able to see and help all this happen. Again, the time parameters have to be appreciated. Often, it took months or years before the individual found his way, but, with support, he finally did reach a goal and continue on a higher, most satisfying level.

In a statistical study, as in the literature, there may be a discouraging report at one or a few years after surgery. More time may be required for changing one's life path. The therapist must provide a context in which the patient feels he has back up for his trials and encouragement if there are failures. Altogether, caring for the epileptic patient is a very parental activity, requiring patience, optimism, continuous upgrading of evaluation, appropriate guidance, and unending availability.

These data should counteract the often pessimistic input in the literature, particularly as to the incidence of psychiatric difficulties after surgery and paucity of success. If one knows one's patients before surgery and follows them in the aftermath, one can serve as an engine for change and negative occurrences can be dissected and useful maneuvers initiated.

The team approach is the vital component, allowing for precise analysis of events and correct action by the most appropriate team member, all possible through continuous contact by each with the patient and communication between the members.

This work may also serve as a commentary on the recognition of the scientific value of a phenomenological approach. To attempt to understand a person's life with objective tests and scales, given at fixed and infrequent intervals, is reminiscent of a comment of Dr. Rayport's that the "objective" approach in assessing behavior is like trying to evaluate the meaning of a Van Gogh painting by counting the number of times blue, red, or yellow appears.

References

Dodrill, C. B., Batzel, L. W., Queisser, H. R., and Temkin, N. R. (1980). An objective method for the assessment of psychological and social problems among epileptics. *Epilepsia* **21,** 123–135.

Duncan, J. S., and Sagar, H. J. (1987). Seizure characteristics, pathology and outcome. *Neurology* **37,** 405–409.

Falconer, M. (1973). Reversibility by temporal lobe resection of the behavioral abnormalities of temporal-lobe epilepsy. *N. Engl. J. Med.* **289,** 451–455.

Ferguson, S. M., and Rayport, M. (1965). The adjustment to living without epilepsy. *J. Nerv. Ment. Dis.* **140,** 26–37.

Ferguson, S. M., and Rayport, M. (1988). A multidimensional approach to the understanding and management of behavior disturbance in epilepsy. *In* "Modern Perspectives in Clinical Psychiatry" (J. G. Howells, Ed.), pp. 302–330. Brunner/Mazel, New York.

Ferguson, S. M., Rayport, M., Gardner, R. G., Kass, W., Weiner, H. G., and Reiser, M. F. (1969). Similarities in mental content of psychotic states, spontaneous seizures, dreams and responses to electrical brain stimulation in patients with temporal lobe epilepsy. *Psychosom. Med.* **3,** 479–498.

Ferguson, S. M., Rayport, M., and Corrie, W. S. (1986). Brain correlates of aggressive behavior in temporal lobe epilepsy. *In* "The Limbic System: Functional Organization and Clinical Disorders" (B. K. Doane and K. E. Livingston, Eds.), pp. 183–193. Raven Press, New York.

Ferguson, S. M., Rayport, M., Blumer, D. P., Fenwick, P. B. C., and Taylor, D. C. (1993). Postoperative psychiatric changes. *In* "Surgical Treatment of the Epilepsies" (J. Engel, Jr., Ed.), 2nd ed, pp. 649–661. Raven Press, New York.

Horowitz, M. J., and Cohen, F. M. (1968). Temporal lobe epilepsy: Effect of lobectomy on psychosocial functioning. *Epilepsia* **9,** 23–41.

Horowitz, M. J., Cohen, F. M., Skolnikoff, A. Z., and Saunders, F. A. (1970). Psychomotor epilepsy: Rehabilitation after surgical treatment. *J. Nerv. Ment. Dis.* **150,** 274–290.

Horsley, V. (1886). Brain surgery. *Br. Med. J.* **2,** 670–675.

James, I. P. (1960). Temporal lobectomy for psychomotor epilepsy. *J. Ment. Sci.* **106,** 543–558.

Milner, B. (1954). Intellectual function of the temporal lobes. *Psychol. Bull.* **51,** 42–62.

Rasmussen, T. (1979). Cortical resection for medically refractory focal epilepsy: Results, lessons and questions. *In* "Functional Neurosurgery" (T. Rasmussen and R. Marino, Eds.), pp. 253–269. Raven Press, New York.

Rausch, R., McCreary, C., and Crandall, P. H. (1977). Psychosocial functioning following successful surgical relief from seizures. Evidence of prediction from preoperative personality characteristics. *J. Psychosom. Res.* **21,** 141–146.

Rayport, M., and Ferguson, S. M. (1996). Psychiatric evaluation for epilepsy surgery. *In* "The Treatment of Epilepsy" (S. D. Shorvon, F. E. Dreifuss, D. F. Fish, and D. G. T. Thomas, Eds.), pp. 631–648. Blackwell, Oxford.

Rayport, M., and Ferguson, S. M. (2001). Psychosis of epilepsy: An integrated approach. *In* "Psychiatric Aspects of Epilepsy: A Practical Guide to Their Diagnosis and Management" (A. Ettinger and A. Kanner, Eds.), pp. 73–94. Williams and Wilkins, Baltimore.

Rose, K. J., Derry, P. A., Wiebe, S., and McLachlan, R. S. (1996). Determinants of health-related quality of life after temporal lobe epilepsy surgery. *Qual. Life Res.* **5,** 395–402.

Serafetinides, E. A. (1962). The effects of temporal lobectomy in epileptic patients with psychosis. *J. Ment. Sci.* **108,** 584–593.

Serafetinides, E. A. (1965). Aggressiveness in temporal lobe epileptics and its relationship to cerebral dysfunction and environmental factors. *Epilepsia* **6,** 33–42.

Spencer, S. S., Berg, A. T., Vickrey, B. G., Sperling, M. R., Bazil, C. W., Shinnar, S., Langfitt, S. T., Walczak, T. S., Ebrahimi, N., and Frobish, D. (2003). Initial outcomes in the multicenter study of epilepsy surgery. *Neurology* **61**(12), 1680–1685.

Taylor, D. C. (1969). Aggression in epilepsy. *J. Psychosom. Res.* **13,** 229–236.

Taylor, D. C. (1972). Mental state and temporal lobe epilepsy: A correlative account of 100 patients treated surgically. *Epilepsia* **13,** 727–765.

Taylor, D. C. (1975). Factors influencing the occurrence of schizophrenia-like psychosis in patients with temporal lobe epilepsy. *Psychol. Med.* **5,** 249–254.

Taylor, D. C. (1987). Psychiatric and social issues in measuring the input and output of epilepsy surgery. *In* "Surgical Treatment of the Epilepsies" (J. Engel, Jr., Ed.), pp. 485–505. Raven Press, New York.

Vickrey, B. G., Hays, R. D., Graber, J., Rausch, R., Engel, J. Jr., and Brook, R. H. (1992). A health-related quality of life instrument for patients evaluated for epilepsy surgery. *Med. Care* **30,** 299–319.

APPENDIX I

IN MEMORIAM: MARK RAYPORT, M.D. C.M., PH.D., F.A.C.S

An Appreciation by George Ojemann

Reprinted from *Epilepsia* 44(9): 1262–1264, 2003 with permission

Picture taken by Beverly Hall, on Nantucket, September 1997, the occasion was Mark Rayport's 75th birthday.

The generation of physiologically oriented surgical epileptologists trained by Wilder Penfield lost a distinguished member with the passing of Mark Rayport, M.D. C.M., Ph.D., F.A.C.S., on March 4, 2003, at age 80 years. Dr. Rayport survived leukemia in 1982, but then a second leukemia developed in 2002. At the time of his death, Dr. Rayport was Professor Emeritus of Neurological Surgery and Radiology at the Medical College of Ohio. Dr. Rayport pioneered an interdisciplinary approach to the study and treatment of epilepsy. His partner in care of the whole patient was Dr. Shirley Ferguson, a psychiatrist and his wife

DOI: 10.1016/S0074-7742(06)76007-X

for more than 50 years. Before doing so was second nature to epileptologists, Drs. Rayport and Ferguson emphasized quality of life (Rayport *et al.*, 1982, 1986, 1987), the psychiatric consequences of resective epilepsy surgery (Ferguson *et al.*, 1993), and the important relations between psychiatric syndromes and epilepsy (Ferguson and Rayport, 1965, 1988; Ferguson *et al.*, 1969a,b, 1986; Rayport and Ferguson, 1996, 2001). From the beginning of his epilepsy surgery career, Dr. Rayport advocated extensive physiologic evaluation of prospective surgical candidates before surgery. In patients with stereotactically implanted electrodes who underwent brain stimulation, it was possible more accurately to define the epileptogenic tissue, but also to substantiate hypotheses about the intrinsic nature of some of the behavioral problems occurring in patients with temporolimbic epilepsy. Dr. Rayport was among the first to use single neuronal unit recordings in human epilepsy (Rayport, 1961, 1972; Rayport and Waller, 1961, 1967). Subsequently he contributed to the performance of corpus callosal resection, emphasizing the value of a two-stage procedure, the search for operable partial epilepsies after callosotomy (Rayport *et al.*, 1993), and the potential for behavioral and speech complications resulting from hemispheric competition (Ferguson *et al.*, 1985; Rayport *et al.*, 1985).

Dr. Rayport's interests in neurosurgery were not limited to epilepsy. He published on a variety of subjects, including hydrocephalus, for which he developed new shunt designs (Rayport and Reiss, 1969), computed tomographic biopsy and cyst aspiration (Savolaine *et al.*, 1987), use of antibiotics (Frenkel *et al.*, 1985; Rayport, 1946), aspects of head injury (Greenblatt *et al.*, 1989; Rayport and Rayport, 1973), and multicentric tumors (Harris *et al.*, 1980). He recognized the importance of the Talairach and Tournoux stereotaxic atlas for the future of neurosurgery, as well as brain imaging, and made the atlas available to the English-speaking world through his lucid translation (Talairach and Tournoux, 1988). Early in his career, he reported on the effects and treatment of addiction (Rayport, 1954, 1956; Wikler and Rayport, 1954).

Dr. Rayport was born in Kharkov, Ukraine, USSR, in 1922. His mother was a physician, and his father, a lawyer. From 1925 to 1930, the family lived in Berlin, where his father was assigned as a commercial attaché by the Soviet government. With the changing political conditions, the family moved to Brussels. After the German invasion of the Low Countries in the spring of 1940, the family fled Belgium across the French border. They lived in France and Portugal for 6 months until they could obtain visas to immigrate to the United States, arriving November 1940. One benefit of this early multicultural experience was Dr. Rayport's fluency in Russian, German, French, Flemish, and English. He continued his education first at NYU and then at Earlham College, from which he was graduated in 1943. He then worked with Simon Komarov, an intellectual grandson of Pavlov, in the Fells Laboratory in Philadelphia. There Dr. Rayport performed the experiments that formed the basis for the Shay rat model of gastric ulcer pathogenesis (Komarov *et al.*, 1944). Komarov encouraged him to apply to McGill University for medical school; he matriculated in 1944, graduating in 1948. Inspired by lectures at the Montreal Neurological Institute, he developed a passion for the study of the nervous system and decided on a career in neurosurgery. He did an internship and medicine and neurology residencies at Bellevue Hospital (on the Cornell Division) in New York. In his last year at Bellevue, he married his wife, Dr. Ferguson. He then entered the US Public Health Service and was assigned to the Lexington, Kentucky, USPHS Hospital for research and treatment of drug addiction. Dr. Rayport returned to Montreal for neurosurgical residency training at the Montreal Neurological Institute, where he was the last of Wilder Penfield's residents. He obtained his Ph.D. in neurophysiology in the laboratory of Dr. Herbert Jasper. In 1958, he was appointed to the faculty of the Albert Einstein College of Medicine under Dr. Leo Davidoff, then chair of Neurosurgery. During his 10 years there, Dr. Rayport directed a research laboratory, established a multidisciplinary epilepsy surgery program, and was an active and valued teacher. In 1966 and in 1967 through 1968, he took sabbatical leave as Visiting Professor at the Laboratoire de Neurophysiologie Comparée, Faculté des Sciences, Université de Paris, France, with Professors Pierre Buser, Jean Bancaud, and Jean Talairach. In 1968, he organized a Festschrift at the New York Academy of Medicine to honor Dr. Davidoff at his retirement (Rayport, 1968). In 1968, he joined the Mt. Zion Neurological Institute in San Francisco, but after 1 year, was invited to join the faculty of the newly established Medical College of Ohio as Professor and founding Chair of the Department of Neurological Surgery. He played key roles in establishing the Comprehensive Epilepsy Program and the Neurosciences program at the Medical College of Ohio. He was recognized in 1987 as the "Ohio Neurosurgeon of the Year" by the Ohio State Neurosurgical Association. He also served as President of the Toledo Neurological Society, and founded the MCO chapter of AOA. In 2001, he and his wife established the Mark and Shirley Ferguson Rayport Fund for Brain Behavior Research in the Medical College of Ohio Foundation. His efforts on behalf of community service were exemplary. Outside his profession, Dr. Rayport was an avid photographer, traveler, chef, and aficionado of the arts and fine books.

He is survived by his wife, Dr. Shirley Ferguson, now Professor Emerita of Psychiatry at the Medical College of Ohio, and their three children. Stephen Rayport, M.D., Ph.D., is on the Psychiatry faculty at the Columbia University College of Physicians and Surgeons. He and his wife Marcia have two daughters, Ilana and Yael. Jeffrey Rayport, Ph.D., is currently CEO of Marketspace LLC, a busibusiness advisory firm. Jennifer Rayport Rabodzeenko is a partner with her husband in a video production company. They have one son, Blake. Dr. Rayport believed in the care of the whole patient, generous sharing of knowledge, academic excellence, and the beauty of the order of nature. His integrity, loyalty, devotion to family, insightful perspectives, and openness to new ideas will be sorely missed.

References

Ferguson, S. M., and Rayport, M. (1965). The adjustment to living without epilepsy. *J. Nerv. Ment. Dis.* **140,** 26–37.

Ferguson, S. M., and Rayport, M. (1988). A multidimensional approach to the understanding and management of behavior disturbance in epilepsy. *In* "Modern Perspectives in Psychiatry" (J. G. Howells, Ed.), pp. 302–330. Brunner/Mazel, New York.

Ferguson, S. M., Rayport, M., Gardner, R., *et al.* (1969a). Similarities in mental content of psychotic states, spontaneous seizures, dreams, and responses to electrical brain stimulation in patients with temporal lobe epilepsy. *Psychosom. Med.* **31,** 479–498.

Ferguson, S. M., Schwartz, M. L., and Rayport, M. (1969b). Perception of humor in patients with temporal lobe epilepsy: A cartoon test as an indicator of neuropsychological deficit. *Arch. Gen. Psychiatry* **21,** 363–367.

Ferguson, S. M., Rayport, M., and Corrie, W. S. (1985). Neuropsychiatric observations on behavioral consequences of corpus callosum section for seizure control. *In* "Epilepsy and the Corpus Callosum" (A. G. Reeves, Ed.), pp. 501–514. Plenum Press, New York.

Ferguson, S. M., Rayport, M., and Corrie, W. S. (1986). Brain correlates of aggressive behavior in temporal lobe epilepsy. *In* "The Limbic System: Functional Organization and Clinical Disorders" (B. K. Doane and K. E. Livingston, Eds.), pp. 183–193. Raven Press, New York.

Ferguson, S. M., Rayport, M., Blumer, D. P., *et al.* (1993). Postoperative psychiatric changes. *In* "Surgical Treatment of the Epilepsies" (J. Engel, Ed.), 2nd ed., pp. 649–661. Raven Press, New York.

Frenkel, L. D., Patno, K., Rayport, M., *et al.* (1985). Safety and effectiveness of intraventricular gentamicin in CNS infections. *Infect. Med.* **20,** 36–40.

Greenblatt, S. H., Long, C. L., Blakemore, W. S., *et al.* (1989). Catabolic effect of dexamethasone in patients with major head injuries. *J. Parenter. Enteral. Nutr.* **13,** 373–376.

Harris, J. H., Rayport, M., Voeller, K. K. S., *et al.* (1980). Multicentric CNS tumors. *J. Neuropathol. Exp. Neurol.* **39,** 359.

Komarov, S. A., Shay, H., Rayport, M., *et al.* (1944). Some observations on gastric secretion in normal rats. *Gastroenterology* **3,** 406–413.

Rayport, M. (1946). Some notes on the pharmacology of penicillin. *McGill Med. J.* **15,** 103–122.

Rayport, M. (1954). Experience in the management of patients medically addicted to narcotics. *JAMA* **156,** 684–691.

Rayport, M. (1956). Addiction problems in the medical use of narcotic analgesics. *J. Chronic Dis.* **4,** 102–110.

Rayport, M. (1961). Physiological properties of strychnine spike of acute and chronic isolated cortex of cat. *Electroencephalogr. Clin. Neurophysiol.* **13,** 102.

Rayport, M. (1972). Single neurone studies in human epilepsy. *In* "Neurophysiology Studied in Man: Proceedings of a Symposium Held in Paris at the Facultâe des Sciences, 20–22 July 1971" (G. G. Somjen, Ed.), pp. 100–109. Excerpta Medica, Amsterdam.

Rayport, M. (Ed.) (1968). Festschrift in Honor of Leo M. Davidoff, M.D.: Proceedings of the Rudolf Virchow Medical Society in the City of New York, 26 (Suppl.). Ferdinand Berger, Horn, Austria.

Rayport, M., and Ferguson, S. M. (1996). Psychiatric evaluation for epilepsy surgery. *In* "The Treatment of Epilepsy" (S. D. Shorvon, F. E. Dreifuss, and D. F. Fish *et al.*, Eds.), pp. 631–648. Blackwell, Oxford.

Rayport, M., and Ferguson, S. M. (2001). Psychosis of epilepsy: An integrated approach. *In* "Psychiatric Issues in Epilepsy: A Practical Guide to Diagnosis and Treatment" (A. B. Ettinger and A. M. Kanner, Eds.), pp. 73–94. Lippincott Williams & Wilkins, Philadelphia.

Rayport, M., and Rayport, S. G. (1973). Effect of skull rotation on x-ray evidence of displaced midline structures. *J. Neurosurg.* **38,** 180–188.

Rayport, M., and Reiss, J. (1969). Hydrodynamic properties of certain shunt assemblies for the treatment of hydrocephalus. 1. Report of a case of communicating hydrocephalus with increased cerebrospinal fluid production treated by duplication of shunting device. 2. Pressure-flow characteristics of the Spitz-Holter, Pudenz-Heyer, and Cordis-Hakim shunt systems. *J. Neurosurg.* **30,** 455–467.

Rayport, M., and Waller, H. J. (1961). Microelectrode analysis of the human epileptiform spike. *Excerpta Med. Int. Congr. Series* **37,** 14.

Rayport, M., and Waller, H. J. (1967). Technique and results of micro-electrode recording in human epileptogenic foci. *Electroencephalogr. Clin. Neurophysiol.* (Suppl. 25), 143–151.

Rayport, M., Corrie, W. S., and Ferguson, S. M. (1982). Contribution of stereoelectroencephalographic studies to reduction of failure rates of cortical resection for seizure control. *Appl. Neurophysiol.* **45,** 468–470.

Rayport, M., Corrie, W. S., and Ferguson, S. M. (1985). Corpus callosum section for control of clinically and electroencephalographically classified intractable seizures. *In* "Epilepsy and the Corpus Callosum" (A. G. Reeves, Ed.), pp. 329–337. Plenum Press, New York.

Rayport, M., Ferguson, S. M., and Corrie, W. S. (1986). Contributions of cerebral depth recording and electrical stimulation to the clarification of seizure patterns and behavior disturbances in patients with temporal lobe epilepsy. *In* "The Limbic System: Functional Organization and Clinical Disorders" (B. K. Doane and K. E. Livingston, Eds.), pp. 171–182. Raven Press, New York.

Rayport, M., Corrie, W. S., and Ferguson, S. M. (1987). The pre-operative electroclinical evaluation determines postoperative seizure control after temporal lobectomy. *In* "Presurgical Evaluation of Epileptics: Basics, Techniques, Implications" (H. G. Wieser, C. E. Elger, and R. M. Hess, Eds.), pp. 335–336. Springer-Verlag, Berlin.

Rayport, M., Lassiter, A. T., Ferguson, S. M., *et al.* (1993). Change of classification of seizures after corpus callosum section opens opportunities for further surgical treatment of intractable seizures. *Epilepsia* **34**(Suppl. 2), 109.

Savolaine, E. R., Greenblatt, S. H., and Rayport, M. (1987). Computed tomography guided intracranial biopsy and cyst aspiration: Accumulated experience in 60 patients. *J. Comput. Tomogr.* **11,** 221–227.

Talairach, J., and Tournoux, P. (1988). Co-Planar Stereotaxic Atlas of the Human Brain: An Approach to Medical Cerebral Imaging. G. Thieme Verlag, Stuttgart.

Wikler, A., and Rayport, M. (1954). Lower limb reflexes of a "chronic spinal" man in cycles of morphine and methadone addiction. *Arch. Neurol. Psychiatry* **71,** 160–170.

APPENDIX II: CONCEPTUAL FOUNDATIONS OF STUDIES OF PATIENTS UNDERGOING TEMPORAL LOBE SURGERY FOR SEIZURE CONTROL*

Mark Rayport

Department of Neurological Surgery, Medical College of Ohio
Toledo, Ohio 43614, USA
Department of Radiology, Medical College of Ohio
Toledo, Ohio 43614, USA

The basis of Dr. Rayport's meticulous thinking and methodology in his neurosurgical treatment of intractable temporolimbic seizures is illustrated. This material should provide guidance and inspiration to the epilepsy neurosurgeon.

I. Definition of Clinical Neuropsychiatry

We define clinical neuropsychiatry as the psychiatric discipline which combines psychodynamic psychiatry with behavioral neurology and biological psychiatry, addressing human behavior multifactorially in function and longitudinally in time.

*This paper was included in a presentation at the Annual Meeting of the Toledo Neurological Society and Medical College of Ohio (March 3, 1998) and titled "Global Outcome of Temporal Lobe Surgery for Seizure Control."

INTERNATIONAL REVIEW OF
NEUROBIOLOGY, VOL. 76
DOI: 10.1016/S0074-7742(06)76008-1

0074-7742/06 $35.00

II. Neuropsychiatric Epileptology

Neuropsychiatry makes a distinctive contribution to the effectiveness of the epileptologic team in areas essential to global assessment by:

A. Case-by-case evaluation of life profile, personality and higher nervous functions, and of their interactions with the epileptic pathophysiology.
B. Pre- and postoperative patient and family support.
C. Postoperative rehabilitation.

III. Scopes of Neuropsychiatric and Neuropsychological Epileptology

A. The assessments contributed by neuropsychological examination principally address cognitive functions whose cortical substrates the epilepsy neurosurgeon wishes to protect along his path to the medial lobe structures, namely language cortex, or sometimes to avoid excising, namely hippocampus. It appears that neuropsychology assesses the affectless brain. A paradox arises at this point.

B. The epileptogenic structures excised during temporal lobectomy for seizure control (anterior temporal neocortex, amygdala, and hippocampus) are major components of the limbic system, a widely connected, largely noncognitive, integrative cerebral substrate of instinctual drives, affect, and memory, thereby of personality. The diagnosis and clinical management of disturbances in these functions are in the conceptual and methodologic domains of psychiatry.

C. The neuropsychiatric epileptologist explores the nature and sources of the behavioral problems of the epileptic persons whose complex condition fluctuates over time with occurrence of seizures which may alter cognition, mood, and thought patterns, and may interact with the intrapsychic and the interpersonal life. Neuropsychologic tests are administered as a stabilized battery, optimally at a time remote from the target symptom, the seizure.

D. Neuropsychiatric and neuropsychologic epileptology are complementary and only partially overlapping. The comprehensive evaluation of behavior is within the discipline of neuropsychiatry.

IV. Experimental and Therapeutic Brain Ablation Paradigms

A. Therapeutic neurosurgical ablation has the same research potential as the classic ablational paradigm of neurophysiology, both entailing removal of structure and function. However, there are significant differences.

B. In the experimental model, the independent variable is the anatomic site of the ablation. The dependent variable is normal function.

To produce deficitary syndrome in bilateral integrative brain systems, such as the limbic system, experimental ablation may have to be bilaterally symmetrical as outstandingly demonstrated by Klüver and Bucy (1937) in the postablational syndrome which bears their names.

C. In the neurosurgical therapeutic paradigm for epilepsy, the independent variable is abnormal function. The dependent variable is the anatomic site.

The aim of neurosurgical intervention is ablation of the abnormally functioning tissue, concisely characterized by Hughlings Jackson (1873; Jackson, 1958) with the perceptive physiopathologic term, "discharging lesion." Ablation of the *discharging lesion* is the basis of clinical recovery.

D. It follows that those partial epilepsy patients who have been rendered seizure-free by excision of the epileptogenic lesion are an optimal study population for dependable correlations regarding the effects of seizure activity on behavior.

V. Algorithm and Policies of Presurgical Evaluation

A. Entry Criteria

1. Medication resistant partial seizures (PS): failure of at least three major antiseizure drugs (ASDs).
2. Neuroclinical history and examination confirm PS.
3. Neuropsychiatric history and examination.
4. Diagnostic electroencephalograms (EEGs) (including T1 and sphenoidal leads, sleep deprived).
5. Magnetic resonance imaging (MRI) temporal lobe protocol.
6. Clinical classification of seizures.

B. Localization of the Epileptogenic Zone

1. EEG and closed circuit television (CCTV) monitoring for spontaneous seizures.
2. Electro-clinico-radiological localization of seizure onset.
3. Definitive classification of seizures and epilepsy.
4. *If findings show electroclinical concordance:*
   a. Neuropsychology test battery.

b. Intracarotid amobarbital test for memory endpoint and lateralization of language function.
c. Patient eligible for temporal lobe surgery.

5. *Lack of electroclinical concordance is the indication for intracranial monitoring* (e.g., clinical ictal onset precedes ictal onset in electroclinical EEG (EcEEG).

C. Indication for Intracranial Seizure Monitoring

1. Clinical seizure pattern consistently unifocal.
2. Ictal extracranial EEG nonlocalizing.

D. Selection Criteria for Intracranial Electrodes for Seizure Monitoring

1. The design of an intracranial electrode system requires for each patient a carefully developed hypothesis as to the probable site(s) of seizure origin based comprehensively on the presurgical evaluation.

2. Avoidance of a monotechnical electrode tradition of electrode placement in favor of an electrode system specifically designed to address the above hypotheses:

a. Stereotactically guided multicontact depth electrodes (Talairach *et al.*, 1974) for epilepsies of medial temporal, frontal, and parietal cortices.
b. Subdural strips or grids for epilepsies of the superior and lateral brain surfaces.
c. Skull-screw electrodes sited according to 10–20 International System for epidural EEG recording.

3. The implanted electrode system should be large enough to provide:

a. Delimitation of the boundaries of the epileptogenic zone (EZ): Stereoelectroencephalography yields a three-dimensional definition of the EZ.
b. Detection of multifocality.

A small electrode system impedes distinction between a seizure propagated from elsewhere to the focal electrode system and a locally arising partial seizure. All epileptic activity looks focal in a small electrode system. An invasive electrode system must show both the location of the EZ and where it is not.

References

Jackson, J. H. (1873). 'On the anatomical, physiological, and pathological investigation of epilepsies'. in West Riding Lunatic Asylum Medical Reports, Vol. 3, pp. 315–339. *In* "Selected Writings of John Hughlings Jackson, Vol. 1, On Epilepsy and epileptiform convulsions" (J. Taylor, Ed.), 1958, 90–111. Basic Books, New York.

Klüver, H., and Bucy, P. C. (1937). Psychic blindness and other symptoms following bilateral temporal lobectomy in rhesus monkey. *Am. J. Physiol.* **119,** 3452–3453.

Talairach, J., Bancaud, J., Szikla, G., Bonis, A., Geier, S., and Pedrenne, C. (1974). New approach to the neurosurgery of epilepsy. Stereotaxic methodology and therapeutic results. I. Introduction and history. *Neurochirurgie* **20**(1), 1–240, French.

INDEX

H

I

L

M

N

O

P

R

S

T

W

CONTENTS OF RECENT VOLUMES

Volume 39

Volume 40

Volume 41

Volume 42

Volume 43

Volume 44

Volume 45

Volume 47

Volume 48

Volume 49

Volume 50

Volume 50

Volume 52

Volume 53

Volume 54

Volume 55

Volume 56

Volume 57

Volume 58

Volume 59

Volume 60

Volume 61

Volume 62

Volume 63

Volume 64

Volume 65

Volume 66

Volume 67

Volume 68

Volume 69

Volume 70

Volume 71

Volume 72

Volume 73

Volume 74

Volume 75

FSC
www.fsc.org
MIX
Papier | Fördert
gute Waldnutzung
FSC® C083411

Kolibri 360 GmbH
Ferdinand-Jühlke-Straße 7
99095 Erfurt, Deutschland
produktsicherheit@kolibri360.de